MRB
Engineering
Handbook

Also available from ASQC Quality Press

Buying and Supplying Quality
Richard T. Weber and Ross H. Johnson

Procurement Quality Control, Fourth Edition
Edited by James L. Bossert

Supplier Certification: A Continuous Improvement Strategy
Richard A. Maass, John O. Brown, and James L. Bossert

Inspection and Inspection Management
Charles Suntag

To request a complimentary catalog of publications, call 800-248-1946.

MRB Engineering Handbook

Robert C. Noe

ASQC Quality Press
Milwaukee, Wisconsin

MRB Engineering Handbook
Robert C. Noe

Library of Congress Cataloging-in-Publication Data
Noe, Robert C.
    MRB engineering handbook/Robert C. Noe.
        p.   cm.
    Includes bibliographical references and index.
    ISBN 0-87389-199-6 (alk. paper)
        1. Engineering inspection.   2. Manufactures—Defects.   3. Public
    contracts.   I. Title.
    TS156.2.N64   1993
    620' .0044—dc20   93-11421
                                                        CIP

10 9 8 7 6 5 4 3 2 1

ISBN 0-87389-199-6

Acquisitions Editor: Susan Westergard
Production Editor: Annette Wall
Marketing Administrator: Mark Olson
Set in Galliard and Avant Garde by Montgomery Media, Inc.
Cover design by Montgomery Media, Inc.
Printed and bound by BookCrafters, Inc.

ASQC Mission: To facilitate continuous improvement and increase customer satisfaction by identifying, communicating, and promoting the use of quality principles, concepts, and technologies; and thereby be recognized throughout the world as the leading authority on, and champion for, quality.

For a free copy of the ASQC Quality Press Publications Catalog, including ASQC membership information, call 800-248-1946.

Printed in the United States of America

  Printed on acid-free recycled paper

ASQC
Quality Press
611 East Wisconsin Avenue
Milwaukee, Wisconsin 53202

Contents

List of Figures

Figure   Page

1 MRB Organization, Authorization, and Procedures

Purpose of Text

The purpose of this text is to share the author's 41 years of experience in the aircraft manufacturing field with those who are interested in the field of material review engineering. This text concentrates on the procedural and structural aspects of the review of material defects associated with costly and complicated structures. This experience has been related to the production of high performance aircraft, but should be applicable to a wide range of products requiring extensive fabrication, assembly, and the dedication of many individuals of varied disciplines. The numerous references to aircraft are biographical and may give some purpose to the author's words.

MRB Purpose, Membership, Responsibilities

The initials MRB, not commonly known outside of industry, stand for *Material Review Board*, a committee of individuals of various specialties assigned the task of determining what to do with incorrectly made parts, otherwise called nonconforming parts. Since the MRB deals with incorrectly fabricated parts, and no one voluntarily admits that such parts exist, it is understandable that the purpose and functions of the MRB are not publicized even within the confines of the corporation. The purpose of the MRB is to first determine and designate the end use, commonly called the *disposition*, of such defective parts or groups (assemblies) of parts already joined together. A second purpose of the MRB, possibly more important than the first, is to determine, (with assistance as necessary) the root cause of the nonconformance and the necessary steps to prevent it from recurring.

The MRB consists of representatives from three distinct agencies: the company engineering department, the company quality department, and the customer's quality branch. The members of the MRB from the company engineering department are called *MRB engineers* and are appointed from the ranks of the engineering section. Often these individuals have a structural, mechanical design, or stress background and experience. Nonconformances on electronic or avionic parts, or assemblies, however, require a completely different technical background, so some MRB engineers require this background and may be limited to work on these specialties only; either by job assignment or by signature limitation. Companies specializing in other disciplines, such as hydraulics, require MRB engineers with these specialties. There may be many such members depending on the size of the manufacturing operation. The prime responsibility of the MRB engineering member is to determine and then specify the nature of the disposition for the defective parts or assemblies brought to his or her attention.

The members of the MRB from the company quality department are called *MRB quality engineers* and are appointed from the ranks of the quality engineering section. These individuals often have a strong background in company inspection, or may have prior experience in engineering, manufacturing, or tooling. The prime responsibilities of the MRB quality engineering members are to assist the manufacturing departments in the determination of the cause of the nonconformance and then undertake the necessary steps to ensure that the nonconformance does not repeat in later production. In addition, quality members of the MRB must approve or reject the disposition provided by the engineering member of the MRB.

The members of the MRB representing the customer have the responsibility for agreement and approval (or disapproval) of the disposition of the defective parts and the actions specified by the quality member of the MRB to prevent recurrence of the nonconformance, which is called the corrective action. These directions, statements, and agreements are provided by written annotation on the document initially reporting the details of the nonconformance.

Members of the MRB generally do not work together as a team, although they do communicate with each other when questions arise. The disposition (instructions) for the defective parts or assemblies must be provided by the MRB engineer, then concurred with by the quality engineer;

then both the disposition and corrective action statements, as provided by the MRB quality engineer, must be concurred with by the MRB customer representative. The work flow generally proceeds in the same order. The effort starts at the point of defect discovery with the initiation of the nonconformance description document which is then routed to the MRB quality engineer (or sometimes to a repair control activity) for induction into the MRB system via manual or computer tracking. The document is then sent through the MRB quality engineer to the MRB engineer for investigation and disposition, then back to the MRB quality engineer for concurrence and the corrective action summary. It is then sent to the customer representative on the MRB for overall concurrence, and back to the MRB quality engineer for logout and delivery to manufacturing for implementation—an often lengthy lapsed-time process. (See Figure 1.1.)

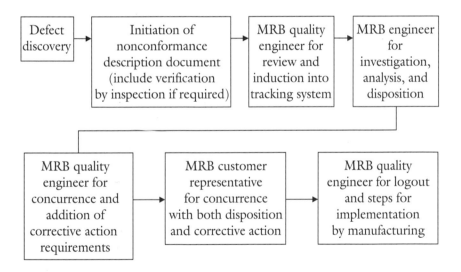

Figure 1.1. Simplified flowchart—material review.

The MRB procedure generally is permitted by the contract, the alternative being the implied scrappage and replacement of any parts discovered defective; a procedure acceptable in some instances, but not where many defects are discovered in permanently joined subassemblies or assemblies of extremely high dollar value. The specifications permitting MRB actions, however, are written in such a way that the procedures (which must be approved by the customer) are considered a privilege, not a requirement, and can be invalidated at any time by the customer.

The Standard Repair

An adjunct to the MRB procedure, primarily to save time, is the procedure for the use of a standard repair. In many instances, similar or identical types of minor defects recur over time on different parts in a random manner. These types of defects can be expected to recur regardless of the amount of corrective action, training, and so forth, directed at the cause. In some cases the causes may be due to workmanship, in others the defects may be expected to recur because the necessary expenditure of $5 million for new tools cannot be justified, or a loss of production for the year required to fabricate such tools may be unacceptable. This is especially true where low production rates cannot support a massive retooling investment and where redesign cannot be accomplished while adhering to unusually strict weight limits. Consideration of the development and design of a standard repair then arises. The usual procedure is the request for a standard repair by engineering or manufacturing personnel, and the design of an appropriate repair by the cognizant MRB engineer, usually following previously accomplished repairs. The repair description is drafted with allowable damage limits, sketches, and matching repair instructions. After review, editing, and so on, the standard repair is approved by a senior member of the MRB engineering staff, a senior member of the MRB quality engineering staff, and a customer engineering representative. Approval for use of the standard repair on the manufacturing floor usually is given by the manufacturer's inspection department. In some instances, the review and signature of an authorized MRB engineer also is required.

Increasing Need for Material Review

Conditions within the aircraft manufacturing industry at the time of this writing (1993) are substantially different than in the past. Until recently, the period was (with brief exceptions) one of growth and widespread development by many manufacturers. New designs were either solicited or independently proposed; contract awards made; and more advanced, more expensive aircraft manufactured. Older aircraft were retired or sold to second-level users and replaced by newer designs. The occurrences of metal fatigue failure were relatively rare, or at least not well understood since the service lives of earlier aircraft generally were not pushed to increasingly greater limits. In addition, the margins of safety built into these earlier aircraft often were higher than those encountered today, partly through design intention and partly through ignorance or conservatism.

In today's economic climate, with relatively less money available for the extremely expensive development of new aircraft, the push is on to increase the life of existing aircraft. Failures through increased use and retesting are occurring. Inspection techniques and analytical procedures for predicting useful lives are improving and many existing aircraft are now being recycled through a reproduction line for detailed inspection, repair, reinforcement for longer life, and update to newer technologies. Existing defects are being uncovered (some possibly present from the time of original manufacture), new defects are being created with attempts to remove damaged parts or replace undamaged parts with those of improved design, and the usual bundle of defects due to workmanship, insufficient tooling, uncoordinated design revisions, and so on, will exist. Replacement of defective parts with the new correctly made parts will not be as much an option as with new production. Repair sophistication must increase. Greater knowledge of structural load paths and distribution among multipart packups with dozens or hundreds of attaching fasteners will be required. The talents of the MRB engineer, whether a member of the liaison engineering section or associated with the structural design, stress, hydraulics, or avionics department, will become increasingly necessary for satisfactory remanufacture of existing aircraft and initial production of brand new aircraft.

It is the desire to assist in the continuing development of this talent that the author is drawing on his personal experience as an MRB engineer to record in one source the learnings of this experience, which may otherwise be lost. Along with some emphasis on sources of information for the fledgling MRB engineer, this text includes some editorial comment, observations, precautions, and recommendations. Knowledge of such should be beneficial; usage will depend on circumstances. The publication of this text is partly an outgrowth of the author's experience with the compilation of a set of MRB engineering guidelines for use by the MRB engineers within his company.

Economic Benefits and Authorization for Use of Material Review

The development and operation of an MRB at a manufacturing facility selling a product to the government evolves from the recognition by both the seller (manufacturer) and the customer (government) that individual parts and assemblies made up of these parts and permanently joined will experience manufacturing defects due to causes which cannot be completely eliminated. Mistakes will occur, regardless of how extensive the training course for riveters, how ready the supply of properly sharpened drills, how extensive the written instructions explaining how to position individual parts against each other, or how sophisticated the tool to properly locate 375 holes in a large web assembly. In one case involving a drilling tool, the tool was not used for a brief period because it was considered too heavy by the mechanics who were supposed to use it. The resulting unfortunate effects were not readily apparent.

When defects (also called *nonconformances*) occur, it may be more cost effective to scrap and replace the part, especially if the part is of low dollar value and readily removable. However, if the part is costly, or if it is permanently joined to other parts, the replacement option may not be viable if costs are to be contained and schedules met. With the possible exception of certain automated parts, manufacturing experience also reveals that few runs of parts manufactured for the typical, low-production-level aircraft program, and even fewer of the resulting complicated assemblies of these parts, will be made defect-free. Thus, in the real world a technique for repair must be employed.

This technique is enumerated in several military documents which are referenced in the applicable purchase contracts. One of the first such documents is the MIL-Q-9858A, Quality Program Requirements. Section 6.5 titled "Nonconforming Material" requires the contractor to establish and maintain "an effective and positive system for controlling nonconforming material" including repair or rework, positive identification, and the use of suitable holding areas for such material. It also states that the acceptance of nonconforming supplies is a prerogative of the government, that the contractor's procedures must be acceptable to the government and that a monetary adjustment may be involved. The last item is not discussed much at the production-floor level. Perhaps it should be.

More specific direction is supplied by reference to MIL-STD-1520C, Corrective Action and Disposition System for Nonconforming Material. This standard states that the act of offering such material to the government should be an exception and that the consistent offering of nonconformances is indicative of a degradation in the contractor's control over quality. Actually, the wording suggests more of a continuation of quality problems rather than a degradation, unless the defect rate actually has increased. Reference to the table of contents of MIL-STD-1520C (see Appendix A) is very enlightening. Many of the terms used in the MRB world are defined within this document and often reproduced within various subsidiary documents generated by manufacturers to comply with the MIL-Q-9858A requirement to establish and maintain an MRB system.

If a subcontractor's operation is large enough, the subcontractor may apply to the prime contractor for permission to set up his own MRB. This most happens when the subcontractor is a vehicle manufacturer or assembly contractor in his own right, or when he has design as well as manufacturing responsibility for the delivered product. This system also is provided for in MIL-STD-1520C. The subcontractor's candidates for MRB engineering and MRB quality engineering membership and the subcontractor's MRB system require approval by the prime contractor's MRB, and the entire delegation must be approved by the government contracting office for the prime contractor. The government representative at the subcontractor's (seller's) plant is delegated the authority as the government representative on the subcontractor's independent MRB. One condition required for the approval of an independent MRB is the presence of an inspection operation and requirement at the subcontractor's plant (often called source inspection); otherwise an independent MRB cannot be set up.

In order for a subcontractor to have his own (independent) MRB approved by the prime contractor for whom he is manufacturing a product, he must pass a vendor certification review. Generally, this is accomplished by the prime manufacturer's MRB quality personnel who review the subcontractor's quality and material review procedures and program in depth and often visit his manufacturing facilities. This is to ensure that not only his procedures meet the requirements of the sub and prime contracts, but that his personnel are familiar with any specific provisions within the letter of MRB delegation.

Once this delegation is granted, follow-up visits and audits, both from an MRB quality and MRB engineering viewpoint, are conducted at intervals.

If noncompliance is uncovered, the MRB authorization can be modified or withdrawn until corrective steps are agreed to, in place, and effective. To avoid this possibility, most companies undertake supplier and in-house training operations, not only for MRB personnel, but also for manufacturing and inspection people expected to work on the project under contract. These exercises take many forms, from classroom lectures and document handouts to requirements for the certification of manufacturing process personnel and others.

In some cases the MRB engineering or quality people have access to a manual or checklist to aid their daily investigations. The author compiled a collection of memos written about MRB engineering policies and procedures over many years and had updated material from these memos published as a set of MRB engineering guidelines. One slightly disconcerting thing was that it was easier for some engineers to call for information than to look through the guidelines for the answer.

Certain key pages from the "C" issue of MIL-STD-1520 are reprinted as Appendix A and contain the table of contents as well as the definitions and explanations of many terms directly applicable to the material review effort. Precise definitions of these terms seem to generate much controversy between personnel of varying disciplines when arguments of application arise on the production floor. At a minimum, an initial read-through is recommended before proceeding. Knowledge of the location of these terms and definitions may be valuable.

2 MRB Documentation and Revisions

Forms Used

Each individual manufacturer or contractor and each individual subcontractor (sometimes called the seller or vendor by the prime contractor) uses her own forms to document nonconformances. There appear to be no standard forms used throughout a particular industry. The forms used to document a nonconformance considered to require full MRB signoff generally are different from the forms used where the nonconformance is considered suitable for disposition by a standard repair. In some cases, a company may initially document the nonconformance on a form suitable for the implementation of a standard repair and then transfer the information to a more lengthy form if full MRB services are required. The forms used to document defects on vehicles returned from service for either repair or modification, or both, may be different from those used for original manufacture. This proliferation of forms often is confusing, particularly to non-MRB personnel, and the need for additions, changes, corrections, and reissues further complicates the flow of paperwork.

The Preliminary Disposition

The normal engineering disposition for nonconforming parts or assemblies can be accomplished completely with a onetime set of instructions (sometimes called the write-up). Occasionally, however, it becomes necessary to require the physical disassembly of certain parts otherwise permanently joined, to permit a complete evaluation of damage that may underlie the visible structure. In other instances, there may be a suspicion of internal damage, or the visible damage may be incapable of precise measurement without enough information initially available to specify a complete repair. In these circumstances, a temporary or partial disposition (called a *preliminary disposition*) is written by the MRB engineer and then countersigned by the MRB quality engineer and the MRB customer representative, just as for a final disposition. When these instructions are methodized through the use of an accompanying work order, the necessary mechanical disassembly can be accomplished on the production floor and any required inspections, measurements, and so on, can be undertaken. The results or findings are then entered either on the document copy supplied to the production or inspection personnel, or on a separate sheet submitted as an attachment to the nonconforming document. When submitted to the MRB engineer (usually through the cognizant MRB quality engineer), the condition can be reevaluated and a final disposition given by the MRB engineer. On rare occasions, a second or even a third preliminary disposition may be required. All preliminary dispositions, however, run the risk of getting lost in the shop and first-person follow-through by the MRB engineer is highly recommended. Cases have arisen where preliminary dispositions have not been returned to the MRB for many months, resulting in the charge that the MRB engineer has been holding up the show. In some cases, the originating MRB engineer may have been transferred to another job, and/or the affected assembly moved to a location and position of less repair accessibility from that when the preliminary disposition was first issued.

Incomplete Information, Use of a Buck Slip

Another technique used to obtain more information than available from the original submittal of a nonconformance reporting document is the use of a *buck slip* or document return tag. This tag is attached to the nonconformance reporting form when it is returned for more information of the type not requiring the physical disassembly of detail parts.

The type of information for which the buck slip is used is missing data such as part numbers, readily obtainable dimensions from the inspection department, and sketches necessary to show specific locations of defects. Conceivably this information might be obtained by the use of a phone call or by personal measurement or sketching by the MRB engineer. Providing this data, however, should be the responsibility of the individuals who are charged with the initiation and preparation of the original document. This would be the mechanic, the laboratory technician, or another individual who detects the error, or the cognizant inspector.

The common incompleteness of the technical information given on the original submittal of the typical nonconformance reporting document to the MRB is a major and universal problem with the system. The training and experience of the initial preparer of these documents often is not of the type required to properly prepare technical information adequate to the needs of the MRB engineer.

The best preparers of nonconformance reporting documents are either the senior inspection personnel (particularly those who have worked closely with MRB engineers) or design engineers on special assignment to assist in the preparation of these documents. It is too much to expect the typical mechanic, versed in the mechanical skills of drilling holes, bucking rivets, and trimming the edges of sheet metal parts, to also be able to detail in writing the necessary information to allow the MRB engineer to analyze the problem. The engineer must be able to design a suitable repair without physically viewing the defect. Since any subsequent review of an MRB engineering disposition often must be accomplished without the actual repaired article available for examination, the description of the defect or nonconformance should be adequate to permit an in-depth review with reference only to the written documentation itself, along with copies of any necessary drawings and specifications.

Required Changes

Nonconformance documents with full MRB signoff (engineering, quality control, and customer) on a final disposition are considered complete. Changes for whatever reason, whether to correct an erroneous disposition or address mechanical or other errors in the attempt to accomplish the disposition, must be undertaken with an official formal change to the original document. This is brought about by the initiation of an A (or B, or C) change. The individual who determines that a change is required must personally initiate or request a supervisor or member of the MRB to initiate a request for an A (or later) change. On the page where the need for a change is written, the initiator includes the specific reason for the change, such as a required revision to the disposition, the discovery of additional defects in the same area as the original, or an error in the reported dimensions. The document is then resubmitted to the MRB for additional consideration and requires a revised, new, or repeat of the original disposition and the full signatures of the three-member MRB, although not necessarily the original three members. It should be noted that the use of a preliminary disposition or a buck slip returned for more information can apply to nonconformance-type document changes, as well as to the initially submitted documents.

Aerodynamic-Type Defects

The documentation and dispositioning of aerodynamic-type defects such as air passage dents, bumps, steps, and gaps between abutting parts, as well as contour hollows and waviness, can be handled in at least two ways. On some programs, each defect or group of defects on a completed panel or major assembly is described on a separate nonconformance document as the defects become apparent and are inspected. On other programs, at the completion of the first assembly, a cover sheet document is initiated and the aerodynamic discrepancies against that particular assembly are documented as an attachment to the originating document. When all the assemblies are fully inspected, with the original defects documented and dispositioned, all the pages are collated and issued as a final, fully dispositioned document with full MRB signoff.

3 Definitions Clarified

Standard Repair Restrictions, Limitations, and Modifications

Among the more confusing aspects of the MRB world are the sometimes subtle differences in the meanings of paired definitions where a slight difference in a specific definition would dictate that one approach be followed rather than an alternate approach. For example, classifications of defects can be major or minor with a substantial difference in the approach required to sell off each to the customer. This chapter introduces some common definitions, provides the textbook meanings, and discusses the aspects associated with each.

The concept of the standard repair is discussed in Chapter 1. Those repairs that affect the MRB engineer working the manufacturing floor are specific standard repairs that require the approval for use and signature of an MRB engineer. Those not requiring this signature can be authorized for use by the inspection department without any engineering involvement.

Each separate standard repair (SR) may include a section for restrictions and limitations; some having none, others as many as a dozen. A restriction is a prohibition against use under certain conditions, whereas a limitation concerns the extent of damage for which the repair may be considered. Accompanying these definitions should be one of the most important (and often misused) caveats in the SR manual: "Material review board required where the limitations described in this manual are exceeded." Another prohibition should be that standard repairs not be modified. Any repair that exceeds the repair instructions in the SR manual requires full MRB approval. Much misunderstanding exists regarding these prohibitions, although some may be intentional since people often seek the easy way out, legal or not. Clearly, the easier way to specify a repair is through use of an illegally modified standard repair rather than seeking a custom repair which can be accomplished only through the full MRB, a much more complicated and lengthy process. An SR is preapproved by the engineering, quality, and customer representatives of the MRB and, with the exception of those SRs requiring separate MRB engineering signoff at the time of each intended use, requires no more than inspection department approval. Conversely, full MRB approval requires three MRB signatures for each use and can easily take days rather than minutes for full processing.

Nevertheless, the rules are clear. The specifics of the SR repair details cannot be modified (within the SR format) and the limits for use, as listed within the restrictions and limitations, cannot be relaxed or exceeded. The only way these can be revised is through the use of a nonconformance document requiring the participation and signoff of all three MRB members. Some manufacturers use a material review report (MRR) form to accomplish this. The MRR requires either a totally customized disposition (if a repair is feasible) or direction for the use of a standard repair, either as written in the standard repair manual or with listed modifications. This is the only way a standard repair can be modified unless the SR manual itself permits some other means.

An example of the use of an SR which has the restriction "Not for use on the XYZ Program" would be the disposition on an MRR written against the XYZ Program and stating "repair per SR 76." The use of the MRR permits the full three-member MRB to review and either authorize or reject the repair attempt. Similarly, another SR having the limitation "not for voids longer than two inches" could be directed for use on a void three inches long as the repair disposition on an MRR. In this context, the MRR is the nonconformance document of choice when no SR exists for the specifics of the defect under review.

The presence of restrictions and limitations on certain standard repairs often comes about at the request of one or more of the engineering, quality, or customer MRB representatives responsible for signing off on the particular SR. For example, the engineering representative in consultation with the stress department may feel that only holes enlarged for the installation of repair rivets no larger than 1/32 inch beyond the diameter of the original rivet should be considered for authorization within the SR format. Conversely, any damaged hole that requires a larger than 1/32 inch oversize cleanout hole will require the review and approval of all three members of the MRB for each separate instance of need. The limitation on the SR would state, "SR not permitted (in other words, MRR required) when required cleanout hole is more than 1/32 inch beyond blueprint diameter."

A restriction on an electrical-type SR stating, "Not for use on ABC Program," would be applied where the specific repair wire splice components are not compatible with the wiring requirements on that program. A limitation of "No more than four individual adhesive voids between honeycomb core and adjoining skin" may have been added as a quality constraint by either the quality or customer MRB representative at the time of the original SR signoff, even though the repair may be considered to restore 100 percent of the structural integrity (strength) to the repaired part. Many arguments have arisen over the precise number of defects allowed, especially when the actual panel at the time of the argument may have five voids and would therefore require MRR rather than SR authorization for repair. Someone is bound to ask "Why can't we use the SR for five separate two-inch-long voids even though the SR permits up to four separate 2 1/2-inch-long voids (for a total of 10 inches of voids in each case)?" A good question, but one that could be repeated, another time for 10 separate one-inch-long voids, and so forth. The limitation of four may have been chosen as an easily applied quality limit based on past statistical data suggesting that few actual panels sustained more than four voids. Many panels may have sustained only two or three voids and the choice of four would have permitted the SR repair of 90 percent of future panels manufactured using the existing technology and tools. Arbitrary perhaps, but an extremely usable technique and one that can be revised as circumstances change over time. The elimination of such restrictions and limitations would lead to a total loss of control over the repair world by the MRB; the ultimate result being the revocation of the standard repair system, a loss that no one would recommend.

Classification As Major or Minor Waiver

The MRR type forms contain a box with two spaces, one denoted major and the other minor. This indicates the classification of the document as describing one or more defects or nonconformances adversely affecting the integrity of the end product when identified as major and not adversely affecting the integrity of the end product when denoted as minor.

In the definition of a minor nonconformance, the key word is adversely. For example, virtually every nonconformance has an effect on the performance of an aircraft, even though this effect may not be amenable to calculation. The loss of .001 inch of material from the length of a 12-foot-long part (miniscule as it may be), could shorten the life of the part or reduce its strength by a small amount. It is an adverse effect to worry about and this often can be estimated initially by engineering judgment alone, with follow-up calculations to affirm it. The philosophy of a nonadverse effect on integrity as a criterion for the definition of a minor nonconformance is well chosen, leaving the definition of a major/critical nonconformance as "other than minor" as a nonconformance bringing about an adverse (hence generally not acceptable) effect on integrity.

The distinction between a deviation and a waiver is important because a disposition (instruction) provided by a duly constituted MRB for the use, by whatever means, of a nonconforming part is a waiver. Quoting from MIL-STD-481B Configuration Control-Engineering Changes (Short Form), Deviations, and Waivers, a *waiver* is "a written authorization to accept an item which, during manufacture or after having been submitted for inspection, is found to depart from specified requirements, but nevertheless is considered suitable for us 'as is' or after repair by an approved method."

A *deviation*, however, is a written authorization granted prior to the manufacture of an item to depart from a particular drawing or specification requirement for a specific number of units or time and without the need for a drawing or specification change.

The approval of an MRR by the full three-member MRB for a defect and associated disposition classified as minor, results in the issue of this document as a minor waiver (not a deviation). Probably 99.9 percent of all defects documented as MRRs or their equivalent are in this category, since in all these instances the MRB can avoid the type of disposition that would cause the nonconforming material (detail parts, subassemblies, assemblies, or installations) to be identified as requiring a major waiver before permission to use is granted. For example, over an extended period of time many individual (detail) parts are dispositioned as scrap by the MRB engineer and these dispositions are not disputed by the other MRB members. In most cases, the parts were scrapped because their use would have adversely affected the mission of the end product if they had been used. The calculations to show their margin of safety may have revealed the resulting margin to be unacceptably negative and a repair may have been impossible to accomplish. Another example may have been that the reported defect was so extreme that the part could not be made to fit against adjoining parts unless a repair action more costly than the procurement of a correctly made replacement part could be undertaken. In these instances, the need for a major waiver was avoided by scrapping the defective part or repairing it in such a manner that the missing strength was restored.

On rare occasions such a defect may be discovered on a part already in place on an assembly worth millions of dollars and buried deeply enough such that the removal and replacement of the part cannot be physically accomplished without completely destroying the entire assembly. In this case, the only recourse is to classify the MRR as major and call for help. If no one within the manufacturer's organization can suggest a solution to this problem, the decision (generally at the program level or higher) will be made to go for a major waiver. This may entail financial penalties, the issue of design engineering identification documentation, the manufacture of backup or spare parts, or all of these. Clearly, this is a procedure to be avoided, but once the decision is made the next step is to prepare a Request for Deviation/Waiver (DD Form 1694 for military contracts). Generally, this is done by personnel within the manufacturer's engineering specification or technical directives group and submitted to the customer procurement contracting officer, through the cognizant administrative contracting officer. Once approved, the recommended disposition of parts can be authorized to the manufacturing level and the instructions performed. The frequency of major waivers is so low that few MRB engineers ever have been associated with one.

Repair Rather than Rework

A distinction between the terms repair and rework is important as the two sometimes are interchanged. A *repair* is the application by direction of a special technique or process, different from the regularly stipulated techniques, for the manufacture of a sound part and resulting in a physical configuration or end product different from that of a correctly processed part or assembly. In other words, a procedure giving a result other than that intended, even though it may be totally acceptable.

A *rework* is a procedure applied to a nonconforming part that will return it to (if once achieved) or bring it up to (if never achieved) the precise configuration or performance intended by the original design. The best example of a rework would be a plate requiring six rivets, but containing only five because the riveter went to lunch before finishing the job. After lunch the inspector, who took a shorter lunch break, picked up the part and wrote up a nonconformance document stating that the blueprint (drawing) required six rivets whereas the plate had only five. The required rework would be for the riveter to install the missing rivet at the correct location. If the plate had seven rivets installed, it could not be reworked back to the required six because, even if the unwanted rivet could be removed, it would still have an unwanted hole.

Occasionally a nonconforming part can be brought to a correct configuration by using a special processing technique. If this special technique is not allowed by the existing specifications applicable to the manufacture of a correct part, the required work is a repair. If it is permitted by the specification, the required work is a rework. Returning to the six-rivet plate example, if the drawing stated "welding permissible for holes or cavities less than 1/4 inch in diameter," the seventh hole, if smaller than 1/4 inch in diameter, could be welded shut and the part considered reworkable. If not, then a designated repair would be required.

Another example of a repair is the designation of a special heat treatment operation to ensure the final hardness (temper) of a steel fitting; the special treatment not being an allowable part of the heat-treat specification called out on the engineering drawing. Sometimes the terms *rework* and *repair* are used interchangeably (although incorrectly) by engineering, tooling, quality control, and manufacturing personnel. Since actual mechanical, electrical, or other type of work generally is done in accordance with specific work instructions, the misuse of these terms seldom results in a mechanical or physical problem.

Interchangeable or Replaceable

Government specification MIL-I-8500 (along with MIL-D-8706 on United States Navy programs) establishes interchangeability and replaceability requirements. *Interchangeable* parts are capable of physical interchange with only the removal of the attachments such as bolts, nuts, screws, and so forth, required to permit replacement. Cutting, filing, drilling, reaming, hammering, bending, or prying should not be required.

In other words, remove the screws that attach the part to its mounting structure, lift off the part and set it aside, then pick up the replacement part, lay it in place against the same mounting structure, and attach it with the same or identical screws. No tweaking is permitted. The only tool required is the wrench or screwdriver necessary to remove the original part. Implied in the definition is the understanding that not only must the interchangeable parts themselves be able to be installed against an individual mounting structure, but that other mounting structures than the original one be similarly interchangeable with each other.

Thus, consider the case of a typical interchangeable panel slated for installation against a portion of an aircraft, where the panel is to be held in place with 24 screws. The interchangeability can be demonstrated by successfully attaching (with the same 24 screws) panels 1, 2, and 3 against aircraft one (each in succession); panels 1, 2, and 3 against aircraft two; and panels 1, 2, and 3 against aircraft three. This obviously requires that careful and controlled attention be given not only to the manufacture of the three panels themselves, but also to the three separate aircraft panel mounting provisions.

The exercises necessary to demonstrate interchangeability cannot take place after completion of the first parts, but must await the completion of additional parts. Details of the demonstrations considered acceptable must be agreeable to both the manufacturer and the customer and may be written into the contract or subject to later negotiation.

Replaceable parts are the same as interchangeable parts (in other words, no fussing around with the means of attachment), except that additional filing, drilling, trimming, and so on, is permitted to accomplish a satisfactory installation. Thus, any missing attachment holes may be drilled, or interferences along the edge of the replaceable panel may be eliminated by trimming off with a file the areas of interference.

The same degree of attention to attachment hole patterns, sizes, and locations is required for the mounting structure holes, but the replaceable parts may be left blank (undrilled) and the edges of their skins or profiles may be left full to permit precise matching to the mounting structure at the time of installation.

When called upon to provide a disposition for a nonconformance against an interchangeable or replaceable part, the MRB engineer must first be assured that interchangeability or replaceability is indeed a requirement for the part. Often this is indicated by the existence of an interchangeability or replaceability note in the notes column, the field of the engineering drawing, or by the presence of the part identification number on an official list of interchangeable/replaceable parts.

Sometimes the requirement for interchangeability or replaceability does not take effect until the manufacture of a later unit beyond an initial production run (or release) of parts, thus giving the manufacturer time to perfect, modify, or proof-out the tooling. Machined parts containing multiple mounting or installation holes, manufactured to close tolerances, intended to be replaced in service at intervals, or sold as spare parts often are considered by implication to be either interchangeable or replaceable. The task of the MRB engineer is to determine if this is so, based on the anticipated conditions of use, and to consider the effects of any defect or nonconformance on the maintenance of these requirements.

A repair designed to restore the structural integrity (strength and life span) of a defective part, but alters the location of one or more holes within an interchangeable hole pattern, may be acceptable at the time of manufacture, but will be returned after the eventual user attempts to install it. In this case, the only recourse would have been to scrap the part or restrict its use to a special end product.

Fatigue Critical, Fracture Critical, and Other Classifications

The concept of fatigue criticality as applied to the design and manufacture of parts probably dates back to the mid-1970s. At that time it had become apparent that more than the usual emphasis should be applied to the design, manufacture, and inspection of certain aircraft component parts.

Fatigue-type failures, although dating back to the early days of the industrial revolution, had not been fully understood. It was not until the failure of a jet-propelled commercial aircraft that the attention warranted for this phenomenon began to be applied. It also had become apparent that the common sampling methods of inspection (whereby only a statistical sample of all the parts made in a particular run or group of the same type parts was inspected) was not adequate to ensure the complete integrity of all the parts made. Additionally, it was apparent that the economic costs necessary to inspect 100 percent of all parts made for each aircraft for compliance with the engineering drawings and specifications would be prohibitive.

Working with government personnel, at least one manufacture set up a system whereby certain parts destined for use on current aircraft in production were to be given special attention and handling beyond the usual techniques customarily applied to such parts. This led to the development of four sensitivity characteristics applicable to certain manufactured parts. These classifications were critical, fatigue critical, interface, and fracture critical.

If a *critical* characteristic was discrepant, it could result in a hazardous or unsafe condition for individuals using or maintaining the product, or in the loss of an aircraft.

An *interface* characteristic generally was a dimension that, if discrepant, would prevent normal assembly and could seriously impact interchangeability, schedule, or cost.

The *fatigue critical* characteristic was used to identify an individual part or assembly of parts, or that particular area of a part or assembly where the stress level was sufficiently high that, if a defect were to occur, a fatigue failure could take place and possibly result in the loss of an aircraft.

The *fracture critical* characteristic was applicable primarily to Air Force vehicles and identified those zones of a particular part where a crack or other defect larger than that defined by specification might result in premature failure of the part with probable loss of the vehicle or major subsystem.

All parts containing any of these classifications required 100 percent inspection of the geographical or other features so identified. No sampling was permitted. The individual features themselves (such as dimensions) were given special letter identification on the applicable engineering drawings, fabrication and assembly instructions, and tools. The responsibility for the selection and identification of critical and interface sensitive parts rested with structural design; while for fatigue critical and fracture critical parts it rested with the stress department. In practice each individual program requiring the use of these classifications had to identify those parts or areas meriting such classification, either by publishing lists or by incorporation of the requisite symbols and notes on the engineering drawings involved.

Additional Fatigue or Fracture Critical Review Required

The handling of nonconformance against parts classified critical or interface required no additional attention beyond that provided by the members of the MRB, except that the engineering member might evaluate the effects of the nonconformance to greater depth than otherwise and design the disposition accordingly. Nonconformances identified against fatigue critical or fracture critical parts, however, required an additional layer of review prior to the determination and release of the MRB engineering disposition.

Fracture control activities generally are the responsibility of a specific group of people, such as a Fracture Control Board set up for this purpose. The MRB engineering dispositions against all parts or assemblies designated fracture critical may require the countersignature of a designated member of the Fracture Control Board. This individual, usually a senior member of the board from the stress department, must assure that any necessary fracture analysis has been satisfactorily accomplished. This approval is necessary before the nonconformance reporting document could be submitted to the quality and customer members of the MRB for their signatures. Generally, a part may be designated as fracture critical or fatigue critical, but not both. Fatigue considerations are based on the shortening effect on the service life of a part containing an unwanted defect of a type that serves to increase the stress applied to the part at the defect's location. Calculations taking into account the revised geometry of the part and the mix of varying loads to which the part is expected to be subjected during service can be undertaken to predict the actual reduced life of the part having the defect.

Fracture considerations take into account the initiation of a crack in a particular part and the rate at which the size of the crack is presumed to increase to the point where this increase changes from a slow, stable rate of growth, to a rapid extension to the point of failure. The resistance of a particular type of material to this change from slow to rapid crack growth is called *fracture toughness*. Calculations can be undertaken to predict the service life of a part using the analytical techniques of fracture mechanics. Similarities exist between the techniques used to predict service life from fatigue and fracture standpoints but they represent different philosophies of approach.

The additional layer of engineering review required when a defective part or area of a part has been designated fatigue critical depends on the effect of the defect in reducing the part's calculated service life. Many types of nonconformances will result in a reduction in the part's anticipated life (hours of use, number of catapults or arrested landings, cycles to failure, and so on). It is the responsibility of the MRB engineer to consider this possibility when reviewing the seriousness of the defect and providing the disposition for use, whether or not the part has been designated fatigue critical. The additional engineering approval, however, may not be officially required unless the part had been formally designated fatigue critical; this applies regardless of the type of defect. Application of an incorrect shade of paint requires this added review and signoff just as much as the discovery of a 12-inch-long crack, once the part is on the fatigue critical list, or is so marked on the blueprint.

Conformance to the need for a fatigue review may first be indicated on the typical nonconformance reporting document by the reviewing MRB engineer. The engineer initials a fatigue critical box on the document, indicating YES if the area of the part against which a nonconformance has been written had been designated fatigue critical, or NO if it had not.

Once the YES portion of the box had been initialed, the MRB engineer (or when necessary a stress consultant) must consider the effect of the nonconformance on the part's life, without any repair or reinforcement if it is hoped that the part can be used as is, or with a repair in place if this is necessary to permit the use of the part rather than scrapping it. This evaluation requires that the reviewing engineer either have or obtain some knowledge regarding the design life of a correctly manufactured part. Since that area had been designated fatigue critical, it is reasonable to assume that a fatigue analysis had been accomplished by the design stress analysts and is available either in the form of a published stress report or in the stress department files of unpublished analyses.

In any case, whether or not a specific analysis had been accomplished for the area under consideration, the applied loads should be available or can be derived by the design stress analyst currently assigned to the program. In some instances this can be done by an MRB engineer who has a stress background. With either a copy of the original analysis, or knowledge of the current design loads, the reviewing engineer should be able to determine (or have determined in consultation with the stress department or an MRB engineer with stress/fatigue analysis capabilities) the calculated life of the part as designed and the calculated life of the part (at the discrepant area), with the reduction in life due to the defect's presence. In general, any mechanical type defect in a fatigue critical area will cause a reduction in the life of the part or assembly. Among these defects are unwanted holes, enlarged holes, mislocated holes either too close to other holes or to a free edge of the part, scratches in the surface of the part, cracks, missing material, and other differences from the configuration of the part as originally designed, all of which result in what is called a *stress raiser*. (See Figure 3.1.)

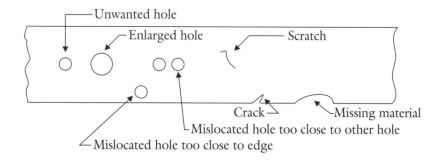

Figure 3.1.   Types of defects causing part life reduction.

Stress is the physical load applied against each unit of area of the part, pounds against each square inch of area, commonly termed pounds per square inch or psi. Since the fatigue life of a part is related to the stress, any change to the configuration of the part causing an increase in stress (a stress raiser) may bring about a reduction in the part's life.

Major or Minor Fatigue Impact, Stress Concentration Factor, Scatter Factor

The degree of additional engineering approval required once a fatigue critical area of a part has been documented as having a nonconformance is dependent on the recalculated life of the part as intended for use. If the part is dispositioned as scrap, no further calculation is required. If the part is dispositioned for use-as-is, the life must be calculated for the physical configuration of the part as damaged or otherwise nonconforming. If the part requires a reinforcement to permit its use (the part may be too expensive to replace, but capable of repair), the life must be calculated with the reinforcement considered to be in place. It is the amount of reduction in the design fatigue life and the actual recalculated life itself that determines whether the change in fatigue life is considered major or minor. If the effect is minor, the nonconformance may be considered to have a minor fatigue impact. If the effect is major, the nonconformance is considered to have a major fatigue impact. Different terms may be in use in different companies.

A minor fatigue impact might be considered to occur when the analysis shows both a minimal increase in the design stress level or stress concentration factor and a fatigue scatter factor of 4.0 or more. A major fatigue impact would occur for all other conditions. The meaning of stress has been briefly discussed and stress level means about the same, but without the precision of an actual calculated stress. Stresses of both 34,498 psi and 35,106 psi would be at the 35,000 psi or 35 kips per square inch (ksi) stress level. (A kip is equal to 1,000 pounds.)

The words stress concentration factor or stress concentration and scatter factor require some explanation. The presence of a discontinuity in the shape or size of an otherwise uniformly shaped part causes a stress raiser. The local stress at the location of the discontinuity has been determined by both analytical (that is, by analysis or calculation) and empirical (based on testing or observation) means to be larger than the stress at the same location if the discontinuity (such as an unwanted hole) had not been present. The ratio between the larger (raised) stress and the smaller stress, given a numeric value, is known as the stress concentration factor K. It is due to the concentrating effect of the stress raiser or discontinuity. Thus, a nonconforming part with an unwanted hole may have a stress at the edge of the hole of 30,000 psi whereas the same part without the hole might have a stress of 10,000 psi. The stress concentration factor due to the presence of the hole would thus be 30,000/10,000 or 3.0, a K of 3.0.

For fatigue life calculation purposes one must know the value of the raised stress, equal to 10,000 psi (no stress concentration factor) on a correctly manufactured part, but 30,000 psi (10,000 x 3.0) on the nonconforming part. Either an increase in the original stress, perhaps due to an increase in the applied load, or an increase due to the presence of the stress raiser will result in a reduction of fatigue life. A minimal increase in the stress level, or a minimal increase in the stress concentration factor, thus are presumed to result in a minimal decrease in the resulting fatigue life.

The interpretation of the word minimal is left to the engineer. A minimal decrease in life from a starting point of 10,000,000 hours could be considered many more hours than a minimal decrease in life from a starting point of 1000 hours. To the author's knowledge, no attempt to state this in terms of a percentage allowable decrease in life has been attempted.

The term *scatter factor* really is a margin of safety for fatigue life. Similar to the static (nonfatigue) margin of safety, the scatter factor requires knowledge of an allowable base point. For instance, the aircraft purchaser's specification, part of the contract in fulfillment of which the aircraft is made, generally states the required life the vehicle must attain in service (for example, 6000 hours of flight).

The scatter factor denotes the amount by which the tested or calculated life exceeds the contractually required life. Thus, a part of, or an entire aircraft tested without failure to 12,000 hours would have demonstrated a test scatter factor of 12,000/6000 or 2.0. The same part, determined by calculation to have an expected life of 24,000 hours, would have an analytical scatter factor of 24,000/6000 or 4.0.

Scatter factors based on actual test are considered more reliable than those based on calculation. So, a required demonstrated test scatter factor of 2.0 may be acceptable to the customer, whereas an analytical scatter factor of 4.0 may be required with applied to nonconforming material, subject to the review of the MRB.

The designation of a nonconformance on a fatigue-controlled program as having minor fatigue impact or major fatigue impact generally would be the responsibility of an MRB engineer with fatigue calculation capabilities who has been designated an MRB fatigue engineer. This individual may or may not be the same MRB engineer who is reviewing and providing the disposition on the nonconformance document for the defective part. In the absence of an MRB fatigue engineer, the designation of the nonconformance as having major or minor fatigue impact also could be accomplished by an approved program stress engineer.

The actual designation is indicated by the placement of the engineer's initial within the fatigue critical box on the MRR. When the defect has been designated as having a minor fatigue impact, the defect must have caused only a minimal increase in the design stress level and/or the stress concentration factor associated with the defect. Simultaneously, the calculated fatigue life due to the defect must have a scatter factor of no less than 4.0, or other agreed upon value. In this case, the MRR can be submitted to the quality and customer members of the MRB for their concurrence, and a copy of the document sent to the stress department for postaudit as necessary. The final document then can be released to the manufacturing area for physical accomplishment of the disposition.

When the nonconformance has been designated as having a major fatigue impact and before final MRB engineering disposition signoff, it must be submitted to the program stress group leader or other designee for a structural review of the effects of the nonconformance. This review is either an original review and analysis or a perusal of the analysis accomplished by the MRB fatigue engineer. When the scatter factor is verified as, or determined to be, say 4.0 or larger, the fatigue box is initialed by the stress group leader. When the calculated scatter factor is less than 4.0 the fatigue box must be initialed by higher level supervisory/administrative personnel from the stress department. The names of these individuals should be kept current and published in association with a listing of approved MRB fatigue engineers. At the time of stress department signoff, the calculated scatter factor may be noted within the fatigue box as equal to or greater than ($\geq$) 4.0 or less than (<) 4.0. When less than 4.0, the actual scatter factor (SF) also may be given.

Only when these major fatigue impact stress department approvals have been obtained can the MRR disposition be given a final MRB engineering signoff and the document submitted to the other members of the MRB for concurrence and release to the shop.

Fatigue Sensitive

Officially, this rather extensive review is only required for those parts, or areas of parts or assemblies designated fatigue critical. However, many other parts may have design SFs of only slightly above 4.0 and may develop an SF of less than 4.0 or sustain a significant decrease in fatigue life if subject to only minimal manufacturing damage. Such parts can be considered as *fatigue sensitive*, not an official definition, but one well understood by those MRB fatigue and stress department engineers undertaking extensive fatigue analyses on damaged or nonconforming parts and assemblies. The MRB engineers are frequently reminded to be alert to this possibility and to consult with the MRB fatigue engineers or stress department fatigue engineers when even the slightest suspicion arises regarding the potential life-reducing effect of any mechanical nonconformance, whether or not the part has been designated fatigue critical.

4 Nonconformance Types and Descriptions—Metallics

Need for Experience and Motivation

An understanding of the various types of defects that may be encountered by the MRB engineer will go a long way toward a realization of the depth of knowledge required by those who come to have a part in the field of material review. A knowledge of what to look for is most important if defects are to be uncovered, documented, and addressed by members of the MRB. An individual with little experience may pass judgment on the correctness of an individual part or assembly of parts in a far different way than one of greater experience. In addition, the motivation and mental attitude, as well as the alertness, of the individual passing judgment go a long way toward the correct evaluation of parts up for review. An individual who is tired, whose eyesight is poor, who is working in a dimly lighted or a noisy environment, or who may be in a rush to complete an inspection of a part or a review of supporting paperwork (backup documentation) will not detect nonconformances that those working under better conditions will detect. Whether a parts inspector or an MRB engineer, an individual who is not motivated to discover all reasonably detectable defects and considers such discovery an annoyance rather than a measure of his or her capabilities will seldom uncover as many defects as the individual who considers his or her duties a search for truth, an opportunity to better ensure the attainment of top quality in the product with which he or she is associated.

It is not the task of this treatise to buttress a positive motivation in the search for defects, a subject fraught with political and economic overtones, but to broaden the knowledge of types of detectable defects discovered in the accomplishment of tasks associated with material review.

Since the field of aircraft manufacturing encompasses a wide range of technological interests, a review of the various types of defects encountered during the inspection of parts slated for installation in today's aircraft will uncover a large spectrum of technology. However, since the author is mechanically oriented, the specifics of electrical or chemical defects will be covered in less depth. Perhaps a future concentration on electrical defects may be beneficial, but in this field the instructions to remove and replace probably are more common than repair as directed.

Defective Holes

The most common defects or nonconformances are those associated with hole drilling. Of the 80 or so standard repairs within the standard repair manual the author used, probably only six or eight accounted for 85 percent of the traffic, and those involved bad holes. Since the joining mechanisms for most aircraft parts involve fasteners of various types and each fastener of the many thousands used requires matching, precision drilled or reamed holes, it is understandable that even a very small percentage of bad holes will represent a sizable number of defects. Additionally, since fasteners always join at least two (sometimes as many as a dozen) individual parts together, a decision to throw away the parts (each varying in price from a few cents to many thousands of dollars) and start over again seldom is a reasonable option. Also, the demonstrable need to repair rather than replace has brought about the development of special techniques and special fasteners, which will be discussed in detail later.

In addition, defects are defined with varying degrees of precision depending on the capabilities and desire of the individual assigned to writing down the description of the defect. The MRB engineer, however, must know precisely what is wrong with a defective hole, sometimes within .0001 inch. For this reason, rather than use the generic term defective hole, the following definitions should be employed. (Explanations will follow each listing.)

Oversized hole, elongated hole, double hole, figure eight hole, extraneous hole, mislocated hole, uncoordinated hole, bellmouthed hole, barrel-shaped hole, and insufficient edge distance (IED) hole. Defective countersinks, counterbores and spotfaces, generally associated with in-line holes will be discussed separately. The interpretations of each of these definitions may vary from individual to individual within a particular discipline (for example between an assembly inspector and his supervisor) and even more from one discipline to another. The following definitions are the author's and represent the MRB engineering viewpoint. They are concerned with holes intended for the installation of fasteners, and hence round, rather than holes for other purposes which may be required to be oval, square, hexagonal, or more freeform in the case of a weight-saving cutout of material.

An *oversized hole* is one that is still round (a term that has varying meanings), but has been enlarged equally all around the periphery or circumference and has the same centerline as the original hole. The tolerance spread required for the original hole (for example, a hole required to have a diameter of .190–.193 inch has a tolerance spread of .003 inch) should not be exceeded for the oversized hole. Thus, a hole required to have a diameter of .190–.193 inch centered at a particular location but, on inspection, discovered to have a diameter varying between .208–.211 inch at the same location, is considered an oversized hole. A .215–.216 inch hole, with a tolerance spread of only .001 inch also would be oversized. A defective hole of .215–.219 inch, having a tolerance spread of .004 (by the author's definition) cannot be considered a true oversized hole although most inspectors probably would identify it as one. Unfortunately, the term oversized hole is widely used to define almost any hole larger than that originally required, so the MRB engineer has the task of determining exactly the configuration and location of the defective hole.

The .215–.219 inch diameter hole should more reasonably be considered an *elongated hole*. An elongated hole has a larger dimension along the longer axis than the dimension along the shorter axis of a two-axis hole (such as an oval, racetrack, slotted, or elliptical hole) and has a tolerance spread greater than the required hole. Refining the definition even further also can require that the center of the elongated hole be at the same center as the desired hole.

This would suggest that the elongation be experienced equally in both directions along an axis coinciding with the axis of the required hole. If not, if the elongation were unequal, the center of the defective hole would be on a new centerline different from the original and would bring another definition into play, that of hole mislocation.

Two other definitions are used to describe an elongated hole. One is a *double hole* and the other a *figure eight hole*. Both definitions emanate from the sequence used to produce these holes. An elongated hole could be produced during a single drilling operation whereby the metal drill itself was either allowed or forced to wander away from the required centerline during the hole's drilling. This might be the case if a mechanic was drilling the hole freehand (that is, with no support to keep the drill centered), and an attractive person walked down the aisle and diverted the worker's attention. Or perhaps the mechanic was using a loosely clamped drill guide or drill bushing that wobbled back and forth during the drilling.

A double hole or a figure eight hole generally is produced by two separate drilling operations not on precisely the same centerline, such as would occur when two separate parts had individual holes in them that were intended to match but, when clamped together, were mislocated slightly, one to the other. The extension of one of the holes along its own centerline through the underlying part would result in a double or figure eight hole in the underlying part. Where the resulting holes are substantially apart, the appearance would be that of a figure eight, with two small points of material called cusps between the two separately drilled holes. If the holes were even further apart so that they were not touching one another, the condition is sometimes called a pair of *binocular holes*. If the drill was allowed to wander sideways between the two holes the cusps would be cut away and the resulting defective hole probably would be defined as a double or elongated hole rather than as a figure eight hole.

Further complications can arise when the defective hole is caused by three individual drilling operations. (The author experienced this once and identified the hole as a Ballantine hole after the three-ring logo of a departed New York City beer.) When a situation such as this arises it is necessary to determine the approximate centerline of the unusually shaped hole because the goal is to determine what size repair diameter round hole will clean out (cut away) the irregularities around the oblong hole. Unusually shaped holes are seldom left as is. The best way to determine the center of such a hole is to position against it the smallest drill bushing that will completely encompass the defective hole. If the hole is in the nearest part facing the viewer, this is relatively easy. If the hole is on the blind side of a packup of parts or is really several individual overlapping holes not yet drilled all the way through, the problem becomes more complex and may require the use of varying diameter pins and knowledge of the original drill sizes used.

A *mislocated hole* is one whose centerline (however determined) is not at the required position. The hole may be of the correct diameter or enlarged.

The concept of mislocation implies that only the hole under review is drilled, not that someone recognized that the hole was in the wrong place and then picked up her drill motor and drilled the hole at the correct location, thus providing two holes to consider. Under this circumstance the incorrectly located hole would actually be an *extraneous* (extra) *hole*.

Thus, the point in time that the anomaly is discovered and written up bears on whether the defective hole is mislocated or extraneous. The MRB engineer must ask himself or others "Was the intended hole also drilled at the correct location?" If so, there is a correctly located hole and an extraneous hole. If not, there is merely a mislocated hole. If both a correctly located hole and an extraneous hole are relatively far apart (at least not touching one another) this difference in definitions is relatively easy to explain. It becomes complicated when a correctly located hole and an incorrectly located hole touch one another or overlap. This condition often is described as the previously defined elongated hole, double hole, or figure eight hole when, in reality, the double hole is the result, but the extraneous hole is the cause. In all cases, the MRB engineer must determine not only the final condition as it exists, but the sequence of events leading to this condition and the purpose of each of the two holes which, if separately drilled, may not have been the same. Cause and effect often are intermingled in this business, and knowledge of each is necessary if satisfactory repair actions are to be prescribed. Intimate knowledge of the manufacturing processes as well as the intentions of the design engineer are necessary if the MRB engineer is to enjoy a successful profession. Equally important is a knowledge of the measurement (inspection) techniques involved and the ability to gather information relative to the attainment of the necessary knowledge.

A *coordinated hole* is one that exists in a location intended to match the location of a common hole in an adjoining or mating part. Once the parts are joined together, coordinated holes in each of two or more mating parts become a single hole along a common centerline.

This hole generally is intended for the installation of a single fastener securing together the parts in the packup. Thus, an *uncoordinated hole* is one that actually is mislocated with respect to the other hole supposed to lie along the same axis when parts are joined. The use of the term *uncoordinated hole* should be limited to one that is not extended through the mating part(s) yet, but if so would result in a double-hole situation depending on the amount of the miscoordination.

The holes in mating parts may each be properly located, but the parts themselves may be physically mislocated, one to the other. The term uncoordinated hole may really spell out the reason or cause of a later defect, not only the result that will occur when the hole is drilled through all the parts in the stackup. The MRB engineer may properly suspicion the use of the word uncoordinated, as well as coordinated, and verify for himself the true cause of the hole mismatch.

Cases also exist where a hole may be properly located but of substantially varying diameters along its length. *Bellmouthed holes* are easy to spot as long as the enlarged (bellmouth) portion of the hole is on the entrance or readily viewable side of the part in which the hole is located.

When the bellmouth is on the far or blind side of a part, it is difficult to even see, much less measure. The depth of the bellmouth may vary but, by definition, it cannot be through the entire thickness of the part. If so, it would be more reasonable to define the defective hole as a tapered hole, which also presupposes a certain uniformity to the taper.

This is not to confuse a tapered hole that is supposed to be straight with one that is supposed to be tapered by design, as for a family of fasteners called *taper-loks*.

An even more difficult-to-spot defect is the *barrel-shaped hole* of correct diameter at the hole's entrance and exit, but enlarged in the middle (much like a beer barrel).

These holes can result from sloppy hand drilling, a broken drill, or the pickup by the drill bit of a piece of debris which, as the bit is being withdrawn from the hole, may enlarge the hole near its center. This is especially possible where the packup of material being drilled consists of hard steel face sheets and a soft aluminum center sheet which can be more easily scored or gouged by a chip from the steel portion of the sandwich.

Measurement of holes having varying diameters along their length is difficult, requiring special techniques other than the use of the simple plug or blade gages used to measure hole diameters at their entrances or exits. Special gages called ball gages or Swiss micrometers or air gages are required, instruments that measure the spread between surfaces directly in line with a specific point along the length of the gage. Facsimile measurements also can be made using a flexible hole-filling material such as a rubber plug that can be hardened in place and then compressed and withdrawn for measurement outside the hole.

Fastener Hole Edge Distance

Another consideration related to this discussion of defective holes is that of hole edge distance, defined as that distance from the center of a fastener hole to the edge of the part in which the hole is located. Some companies use the term edge margin rather than edge distance, so the MRB engineer must be sure of the precise term to use on the particular program under review. See Figure 4.1.

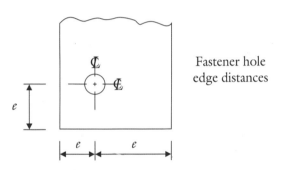

Fastener hole
edge distances

Figure 4.1.   Edge distances.

The edge of concern generally is the nearest edge to the center of the hole, but the words may apply to either the nearest or the farthest edge; the measurement itself being smaller for the edge distance to the nearer edge. The reason for concern about the edge distance is that the strength of a part (and, more important, the anticipated or expected life of the part) is related to its edge distance—a lesser edge distance providing a substantially lesser life than a greater edge distance. So, a concern for edge distance must exist at the same time as a concern for any defects in the hole itself. An oversized hole alone should cause no change in its edge distance, but a mislocated hole will. If the direction of mislocation is toward a nearby edge, the edge distance will be decreased by the amount of the mislocation. It also should be noted that a correctly located hole will have a reduced edge distance if the actual edge of the part in which the hole is located has had some of its material cut away during a previous operation or during its original fabrication.

Again, the cause and effect relationship comes into play. Many a document has been submitted specifying a reduced edge distance due to a mislocated hole when, in fact, the hole was in the correct place and the edge itself was mislocated due to part mislocation before drilling or due to shy material along that edge of a correctly located part.

Insufficient Edge Distance

The term that is employed to describe an edge distance reduced below some defined minimum acceptable dimension is *insufficient edge distance* or, more commonly, IED. Most aircraft parts are designed to have an edge distance equal to twice the diameter of the fastener hole from whose centerline the dimension is measured, commonly called an edge distance (e) of $2D$. (See Figure 4.2)

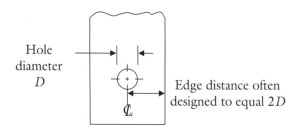

Hole diameter D

Edge distance often designed to equal $2D$

Figure 4.2.   Common design edge distance.

A widely accepted standard of allowable material strengths, MIL-HDBK-5, Metallic Materials and Elements for Aerospace Vehicle Structures, lists bearing strength values for holes having $2D$ edge distances (that is, holes having an edge distance to hole diameter ratio (e/D) of 2) as well as holes having an e/D ratio of 1.5, or 1 1/2 diameters edge distance.

For inspection purposes a generally minimum acceptable edge distance may be given as 1.5 hole diameters, although many examples exist where the actual edge distance required to maintain adequate strength and service life will be more or less than this value. All measured edge distances less than the minimum acceptable values published should be considered as IED nonconformances.

Fastener Hole Spacing and Location

A final consideration in this discussion of hole defects is that of hole spacing, the distance from the centerline of one (round) hole to the centerline of any particular adjacent (round) hole. Sometimes the width of remaining material between two adjacent holes (often called the ligament) is used for structural calculations; this would be a requirement where one or both adjacent holes were not generally round and thus would not have a readily determinable centerline. The term ligament also is used to define the material remaining between the edge of a hole and the edge of the part in which the hole is located.

Hole spacing is listed as a defect when the distance between holes (however defined) is different from the acceptable range of spacings on the engineering drawing. In practice, however (except on any drawings such as those for machined or other close tolerance parts where the required spacings are actually given physical dimensions), little attention is given to the spacings of holes unless they are glaringly too close together or too far apart. This particularly applies where the required locations for the holes are lofted (drawn to full scale) rather than dimensioned. The theory is that if engineering required their locations to be closely controlled, they would have shown some dimensions on the blueprint instead of permitting the tool makers and the inspectors to merely scale the full-scale drawing provided. Rest assured that if engineering later discovers that the hole locations are critical, the necessary controlling dimensions will appear on the drawing.

Fastener Axis Angularity

So far this discussion has covered the types of fastener holes that would be required for protruding head fasteners, requiring no recess or cutting away of material at either end of the hole. Many fasteners, however, are required to be flush with the surface of the part into which they are inserted, generally on one end called the head end or side, but occasionally on both ends or sides.

The word *end* refers to the fastener itself, whereas the word *side* refers to the part into which the fastener is inserted and then installed. Additionally, some fasteners are installed in a packup of material where the material at the tail end of the fastener is not parallel to the surface at the fastener's head end, and either or both surfaces are not the more common 90° to the lengthwise axis of the fastener.

In this case, unless some sort of self-aligning device is used under either the head or tail of the fastener, the sloping surface of material in the area of the fastener head or tail must be cut away to provide a smooth 90° platform surface for the protruding head of a protruding head fastener or for the tail-securing device (nut or collar) of either a protruding head or flush head fastener. When the fastener is similar to a rivet (where the tail is mechanically squeezed, pushed, upset, pulled, or otherwise formed against the surface of the part at the fastener's tail end), the angle of the surface need not be precisely 90° to the fastener's axis since the metal of the fastener will flow against the slope if it isn't too extreme. In this case, the allowable variation from 90° must be determined either by reference to the appropriate fastener or fastener installation specification, or by test program results. Most fasteners with threaded tail or shank, however, use threaded or grooved retainers called nuts or collars, rotated into place along the threads. Except for self-aligning varieties, this requires an abutting surface 90° to the fastener's axis.

These types of fasteners, various types of bolts, screws, pins, and so forth, are widely used throughout the aircraft industry, often in highly loaded areas. In addition, for clearance purposes it may be necessary to recess the nut or collar on the tail end of a threaded-type fastener, even though the surface of the part through which the fastener is installed already is 90° to the fastener axis.

Defective Countersinks

The recess provided below the surface of a part in which a flush head or flush tail fastener is to be installed is called the *countersink* (often abbreviated csk), as is the tool used to cut the recess. The countersink must be in line with the axis of the hole and concentric to it.

Most countersinks are funnel shaped with an inclusive angle of 100°; that is, 50° to either side of the hole axis to match the 100° angle on the head of the flush fastener to be installed. Countersink angles of 78°, 82°, 90°, 130°, and 160° also are used for certain applications.

Defective countersinks are associated with the straight holes of which they are a part; holes which themselves may be correct or separately defective. The most common countersink defects are countersinks too deep or too shallow. Countersinks also may be eccentric to the straight hole axis, oversize (the same as too deep), undersize (too shallow), or elongated. Subverting the definition of an elongated hole as one whose excess enlargement along the major of two axes of enlargement is spread equally from both sides of the required hole, one can expect an elongated countersink to be elongated equally or unequally. The axis of a countersink also may be at an angle to the axis of the hole; that is, the countersink may be tilted. A mislocated countersink must be associated with a mislocated hole, but an extraneous countersink can be either part of a correctly located hole (that is, a hole that should not have been countersunk at all), or part of an extraneous hole that also was countersunk.

When a countersink is too deep, the manufacturer generally is stuck with it because the missing material cannot be restored. However, there may be possibilities along these lines with weldable metals and nonmetallics such as fiberglass and composites having adhesives as part of their structure. A shallow countersink, however, is theoretically reworkable to the required depth by merely countersinking the hole an additional amount. For this to be done, the surface of the part must still be physically accessible for the use of the countersinking tool and there must be no fastener installed, or at least one that can be readily (and economically) removed and replaced. In most cases of a reported shallow countersink, there also is an installed fastener, probably with its flush head unacceptably high and of a type that cannot be easily removed (such as when a wrench cannot be placed on the nut of a flush-type screw). In some cases, an access hole may have to be cut in the underlying structure to permit fastener removal, countersink deepening, and fastener replacement.

Counterbores and Spotfaces

The recess provided below the part's surface when the fastener head or tail must be lowered (or when an otherwise sloping surface must be flat or at 90° to the fastener axis) is called a *spotface* or a *counterbore*. The cutting tools have the same names.

A counterbore is a better choice than a spotface because the depth of a counterbore (or the remaining thickness of the part after counterboring) is controlled by a dimension, whereas a spotface is controlled only to that depth necessary to clean up an irregular surface with little control over the remaining thickness of material. A back counterbore is one applied to the back or far surface of a part. It can be created by putting the tool toward the front face of the part rather than pushing it against the part, which is the more common way of accomplishing a counterbore.

Counterbores are dimensioned with a diameter and a corner or cutter radius, and the depth of cut or the remaining material to be left generally is specified. (See Figure 4.3.)

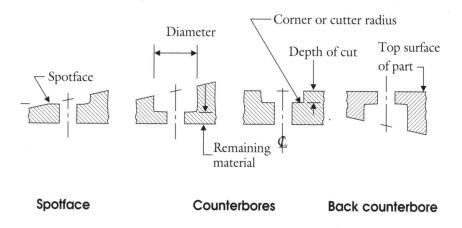

Spotface   **Counterbores**   **Back counterbore**

Figure 4.3.   Spotface and counterbore.

Any or all of these material removal operations can be violated during the attempted accomplishment. Similar to countersinks, counterbores may be too deep, too shallow, eccentric to the underlying hole, or angled to the hole. In general, counterbores (and to a lesser extent countersinks) will be in line with the underlying hole, especially when a piloted tool is used (that is, one that has an attached spindle which guides the tool down into the hole before the cutting of material starts). Of course, the use of a loose or undersized pilot might result in a sloppy counterbore or countersink, especially if the power tool holding the cutting tool is hand held. The author's experience suggests that practically no defect is impossible, over time.

Damaged or Incorrectly Installed Fasteners

Defective fasteners are not often listed as discrepant since they usually can be removed and replaced. When the defect specifically applies to the fastener itself, it is because the fastener cannot be either physically or economically removed and replaced. These defects may range from reports of physical damage to the fastener (such as a cracked head, a drill run into the fastener head, tail, shank, collar, and so on) to an unacceptable position of part of the fastener.

This could result in the fastener head being too high rather than flush (probably due to a shallow countersink); the fastener head or tail hitting a nearby fastener or adjoining part; or the head, tail, or collar of a fastener not lying flat against the surface against which it is supposed to be securely seated. If this is caused by the hole for the fastener being located too close to a leg or extension on the part in which it is installed, the fastener is said to be riding the radius (between the flat and the leg). See p. 206 for a further description.

Head gaps may also be caused by holes drilled at an angle to the surface, or to defective countersinks, spotfaces, or counterbores.

Violations of Part Description Requirements

In addition to hole-type defects written against detail (individual) parts or the assemblies of such parts there are many other types of defects to which detail parts are subject. These defects are not only mechanical in nature, but also affect the metallurgical or physical properties of the material from which the parts are made. Most detail parts are shown on their respective engineering drawings in a form that allows their manufacture to the requirements included on or with the drawing. Thus, the drawing shows a sketch or full-sized representation of the part with as many different views as necessary to permit its fabrication. These views will be either dimensioned or shown full-sized (or both) on a drawing made of material not subject to shrinkage, stretching, or other change in size over time.

Included with, or on, the drawing (or in some cases a full-size mold, casting, or other representation of the part) will be additional information describing the type of material from which the part is to be made, the size of the block of material required for the manufacture of the part (the stock size), the final temper or hardness required, any special processing necessary to fabricate the part, and any final surface coatings or finishes to be applied to the part. Any one of these requirements or procedures can be violated and the task of the inspection process when applied to these parts is to ensure that they have not.

Out-of-Tolerance Dimensions

The most common defects ascribed to detail parts are violations of dimensions including those which define the size and location of fastener holes. Dimensions may be undersized or oversized beyond the tolerance range set up on the drawing for each type of dimension. The tolerance represents the allowable departure or spread from an absolutely perfect measurement (which theoretically can never exist) and permits an acceptable amount of error in either one or both directions from the listed dimension.

Tolerances always are given a directional notation of plus (+) and/or minus (−). The plus and minus tolerances indicate an allowable increase or decrease in the listed dimension. Sometimes the magnitude of the plus tolerance is identical to the magnitude of the minus tolerance. For example, a dimension of 5.312 inches representing the required length of a rod with a tolerance of plus or minus .010 inch (± .010 inch), given as 5.312 ± .010 would accept all rods made between a length of 5.302 inches (5.312 − .010) and 5.322 inches (5.312 + .010). This is an example of a bilateral tolerance which permits a departure from the basic dimension in both directions, although the magnitude in both directions need not be the same for the word bilateral to apply. Another bilateral tolerance would be $5.312 ^{+.012}_{-.015}$.

A unilateral tolerance would be one permitting a departure in one direction only, such as $5.312 ^{+.000}_{-.015}$, actually 5.312 to 5.297. Any dimension having a measurement beyond the dimensional range or spread permitted by the applicable tolerance can be considered defective and the subject for MRB action.

For evaluation purposes, a defective dimension beyond the maximum allowable often is augmented with a number stating the amount by which that measurement is either beyond the largest measurement permitted (over high limit or OHL) or less than the smallest dimension permitted (under low limit or ULL). If the theoretical rod, required to be 5.312 ± .010 (5.302 to 5.322) inches long, measured 5.299 it could be listed as, "Should be 5.312 ± .010, is 5.299 (.003 ULL)." Similarly, if measured as 5.328, it might be listed as, "Should be (S/B) 5.312 ± .010, is 5.328 (.006 OHL).

Any dimension on an engineering drawing is subject to variation and the more dimensions, the more chance for errors. It is not unusual to have dozens or even hundreds of dimensions to describe a complicated part; the more dimensions, the more costly the part, not only to make but also to inspect. It is helpful to realize that the natural tendency for the machinist, when determining the amount of cutting to do on any particular part, will be to aim for the safe side of the tolerance spread and leave the part a little fuller than the minimum dimension. Thus, a modest error can be corrected. You always can remove a little excess metal, but you can't replace missing material without special techniques.

Surface Damage—Gouges, Dents, Cracks

Detail parts, as well as assemblies, are subject to local surface damage. Gouges or indentations can occur on the surface of such parts, either by being accidently hit with a cutting tool or another part, or by being dropped. If the part is thin it may be dented inward, rather than gouged, with the inside face of the part being displaced in the same direction and by the same amount as the outside face; the thickness or distance between the two faces remaining the same. This is an important distinction. The gouge represents an actual loss of material, whereas the dent indicates a displacement or relocation of material. In some cases both may be present and a high degree of observation and measurement is necessary to confirm it.

Parts also can be cracked. This type of defect is among the worst because not only is it very difficult to see, but it may cause an exaggerated reduction in the life of the part, especially if it is subject to loadings that are repeated frequently throughout the part's life. The easiest cracks to detect are those caused by a part being dropped or bumped against another part, workbench, fixture, or floor, especially if the impact causes a visible dent, abrasion, depression, and so forth. In the case of visible damage, it is routine to remove the finishes (surface coatings) from the part and initiate special inspection techniques to ascertain if a crack also exists. These techniques vary from visual magnification to both chemical and electronic techniques to enhance the area of the crack for better observation, either in the area of the crack itself or, via electronic readouts, from instrumentation that is coupled to the testing device. The trick is to know when to conduct the test because the need is not always obvious. In general, however, obvious mechanical misuse of a part must raise the specter of crack possibility. MRB engineering and inspection must work together to determine if resultant cracks exist and, if so, to measure and document their size and precise location. Properly addressed, cracks seldom are left in place beyond a specified interval. (See Figure 4.4 and p. 215.)

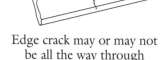
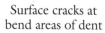

Edge crack may or may not
be all the way through
the thickness

Surface cracks at
bend areas of dent

Figure 4.4.   Surface cracks.

Surface Roughness

Surface roughness is another area of concern for the inspector and MRB engineer and must be carefully evaluated, especially when this roughness may hide underlying cracks or act like a stress raiser. Surface roughness is a measure of the relative smoothness of a surface and, on metallic parts, defines the arithmetic average of the irregularities along the surface from a mean datum. Various types of machining operations and tools produce varying roughnesses of the surface, from a reading of 2000 at an edge formed by a cutting torch, to a reading of 0.5 along a mirrorlike surface. The values represent microinches (.000001 inch) and the reading for a standard machined surface of 125 is indicated by the symbol 125√. The required surface roughness generally is indicated on the engineering drawing and, if defective (usually rougher than required), can be improved if access to the surface itself can be attained.

Surface Finishes

Surface finishes also are subject to defects, both from localized damage due to abrasion and from original misapplication or incorrect processing. If a restoration to the finish requirement is possible and economical, such reworking will be done and the MRB may not be aware of the original defect. If restoration is not possible, and the manufacturing people are looking for an acceptance as is or an easily applied alternate, then the non-conformance is documented and the MRB is alerted. Care must be taken to distinguish between surface finish (applied processes and coatings, such as anodize, alodine, hardcoat, aluminum flame spray, platings, primers, paints, and so forth) and surface roughness (the degree of irregularity of the surfaces's face). One type of surface treatment that seems to lie between surface finish and surface roughness is passivation, a chemical method of restoring desirable oxides to the surface of certain stainless steels after they have been machined, ground, or polished, which disturbs the normal oxide coating on these surfaces. Control of this treatment is provided within MIL-S-5002.

Incorrect Material, Alloy, or Temper

Other defects occasionally experienced by detail parts arise when they are made of the wrong material (for example, aluminum instead of steel), the wrong alloy of the right material (for example, 7075-T6 aluminum instead of 2024-T3 aluminum) or the wrong temper of the right alloy of the right material (for example, 7075-T6 instead of 7075-T73). As in the case of surface roughness and surface finish, any legal rework or removal and replacement of such parts will be accomplished by manufacturing personnel. Only those parts that cannot be economically corrected will be brought to the attention of the MRB. Since routine checking for material content normally is done at the detail part inspection level, only those special cases accidently discovered on the assembly line may merit MRB attention. In one instance a part was discovered to be steel rather than aluminum when the mechanic had to replace dull drills three times while drilling a deep hole in a supposedly aluminum part.

Differences in the materials of detail parts, once installed in an assembly, are more difficult to detect since the part cannot be individually weighed; aluminum, for example, is about one-third the weight of steel. Using various types of hardness testers, checks of surface hardness are revealing, but only give the hardness of the part near the surface. Many steels are magnetic, but certain types of stainless steels are not. Among the more revealing tests are those measuring electrical conductivity, but occasionally a piece of the part under suspicion must be removed and sent to the laboratory for a destructive test and/or chemical composition determination and evaluation. Unless the MRB engineer has become familiar with metallurgical phenomena and can relate to differences in material properties, temper restoration, and so forth, she must consult with metallurgical specialists when confronted with these types of defects.

As noted, the most severe type of metallurgical defect is the use of the wrong material for the manufacture of a part. Wrong materials also may show up in the manufacture of nonmetallic parts, but such incidences are less common. The determination of material content, temper levels, unusual grain structure, and other metallurgical anomalies is a specialized field of investigation. In most cases such investigation can only be done by inspection (quality) personnel specially trained in these techniques and using scientific equipment not available to the average MRB engineer. For example, a spectrographic analysis is required to determine the element content of metal alloys and requires a small piece of the part to be tested. These and other tests, which the MRB engineer should be familiar with, generally require the expertise of a qualified quality control or metallurgical laboratory. Either one should be associated with the company or a competent independent laboratory. A knowledge of the capability of these techniques and others (such as photomicrographics, tensile testing, hardness measuring, and conductivity testing) would materially help the MRB engineer to at least know the proper questions to ask.

Improper Processing

Improper processing (more a cause of defects than a defect itself) may result in a host of individual metallurgical (and nonmetallurgical) defects because most manufactured materials, ranging from chemicals and plastics through various types of metals, are not ready for use in the raw state and must be processed through two or more intermediate steps. It is the possibility of improper processing at one or more steps in the manufacturing cycle that makes the need for process control and the initiation and maintenance of process records so important. Unfortunately, the finished product of an improper process may look the same as one properly processed. An understanding of the types of defects possible through improper processing is very valuable to the MRB engineer. The dropping of a part onto the floor is not considered improper processing even though the manufacture of the part may not be complete. The results of this handling-caused defect and other errors resulting in physical abuse to the part (oversized holes, beyond tolerance dimensions, and so forth) have been commented on previously and are not processing problems. They occur on all types of rigid parts, usually after the processing cycles are complete. For the sake of this discussion, processing problems can be considered as variations from the stipulated parameters of manufacture needed to produce an acceptable final product, but prior to further mechanical actions necessary to revise the shape or configuration of that product. Typical of process actions that can be defective are the control of chemical content, temperature, time, atmosphere, and pressure and/or vacuum. All of these parameters must be controlled between acceptable limits for the entire process to be correct. Indeed, these unwanted variations may occur during the processing or reprocessing of metal alloys, but more common is the selection of the wrong alloy or the wrong temper of the right alloy from which a specific number of individual parts (called a lot, batch, run, or release) has been made. The simple response to this defect is to scrap and replace the parts, but the more economical response might be to consider whether reprocessing or reheat treatment is possible. A knowledge of the processing variables is helpful in this respect, but for specific answers the MRB engineer must, if not metallurgically oriented, consult with one who is. Close ties are encouraged between MRB engineering personnel and metallurgists and material specialists within the company. Much information also is available from material published by the metals industry and by government agencies. See MIL-H-6088, Heat Treatment of Aluminum Alloys for aluminum and MIL-H-6875, Heat Treatment of Steels (Aircraft Practice, Process For) for steel.

Aluminum Alloy Heat Treatment, Temper, and Cold Work

The following text explains in detail the processing required of various metals and alloys, as well as the temper designations applied to them. The information is not given to help the MRB engineer become a metallurgical expert, but to describe the range and scope of the type of processing applicable to metals, any one of which may be subject to error. The material properties achieved by various types of metals are based on processes applied to these metals that are different from one family of metals to another. The treatments applied to aluminum are different from those applied to steel and even vary between the different alloys of each type of metal. For this reason, comments will be made regarding the processing for each family of metals (and non-metals) that apply to that family of materials.

Aluminum alloys (that is, metals consisting primarily of aluminum but containing at least one other element such as copper, zinc, silicon, manganese, or magnesium) are divided into two separate classes. One class' final properties can be varied by differing exposures to heat during the fabrication process and are called heat-treatable aluminum alloys. Those aluminum alloys that cannot have their properties varied in this manner are called non-heat-treatable. The purpose of these treatments is to achieve varying mechanical properties such as strength at low temperatures, strength at high temperatures, resistance to fatigue failure, to fracture, and to environmental corrosion, and stress corrosion. Non-heat-treatable aluminum alloys achieve varying strength levels by the application of different degrees of cold work (that is, physical manipulation such as stretching or compression during the fabrication cycle and sometimes followed by a special thermal (heat) treatment.

The word *temper* is used to designate the physical state in which the aluminum alloy exists and is identified by an alphanumeric code for both the non-heat-treatable and the heat-treatable classes of aluminum alloys. Temper designations for the non-heat-treatable aluminum alloys start with the letter H and are followed by a number 1, 2, or 3 indicating cold working (strain hardening), cold working followed by partial annealing (intermediate heating), or cold working followed by another thermal treatment. Additional digits indicate the specific degree of hardness. Non-heat-treatable aluminum alloys are not as widely used in aircraft structure as are the heat-treatable alloys because they cannot attain the strength-to-weight levels obtainable with many of the heat-treatable aluminum alloys.

Temper designations for heat-treatable aluminum alloys are much more complicated than those for non-heat-treatable aluminum alloys because of the greater variation in processes possible including, not only different types of cold work, but solution heat treatment, precipitation hardening (natural aging or artificial aging) and stress relief by stretching, compressing, or both.

Temper designations for the heat-treatable aluminum alloys start with the letter T and are followed by from one to four numbers, each representing a particular thermal process or mechanical manipulation. For example, T6, a common temper, indicates that the part is to be solution heat treated and then artificially aged, while T8 requires that the part be solution heat treated, cold-worked, and artificially aged. T651 indicates that T6 parts must be stress relieved by stretching before the artificial age cycle. For a full listing of temper designations refer to Chapter 3 of MIL-HDBK-5, or MIL-H-6088.

The term *annealing*, designated as temper 0, is applicable to both non-heat-treatable and heat-treatable aluminum alloys and refers to the heating of the alloy to a temperature within the range of recrystallization (between 600° and 800° F depending on the alloy) and then slow cooling. This removes the effects of all previous cold working or heat treatment and provides a stable and soft (ductile) material suitable for easy forming.

The basic mechanism for strengthening of heat-treatable aluminum alloys is the control of the alloying elements within the solution, particularly controlling the precipitation of the alloying elements from the solution. Solution heat treatment requires the heating of the part to just below the melting point for a long enough time to permit the alloying elements to enter completely into solution, and then rapidly cooling the part by quenching it in water, liquid nitrogen, polyalkylene glycol or other quenchant. This results in a supersaturated solid solution which is unstable at room temperature and is designated by the temper symbol W.

After quenching, the alloying elements within the supersaturated solid solution will precipitate from the solution, the speed of the precipitation depending on the alloy. This results in what is termed *precipitation hardening* or *aging*. If the precipitation takes place rapidly at room temperature (within several days), the term used is *natural aging*, and the temper is considered T3 or T4. When this action is too slow, the part is heated to an elevated but intermediate temperature for a precise period of time to speed up the precipitation process which is called artificial aging, denoted by tempers T5, T6, T7, or T8 depending on what other processes accompany the artificial aging.

Steel Types and Heat Treatment

Steel (a compound consisting of iron and varying amounts of carbon) is alloyed with many other elements and processed in many different ways to produce a wide range of mechanical and physical characteristics. Military Handbook MIL-HDBK-5 lists 23 separate categories of steels ranging from carbon steels, low allow, intermediate alloy, and high alloy steels to precipitation and transformation hardening stainless steels and austenitic stainless steels (far too many to discuss in detail here). Chapter 2 of MIL-HDBK-5, various industry publications, and MIL-H-6875 are valuable sources of information. Some general comments are useful. As for aluminum, steels can be softened by heating and hardened or strengthened (not the same thing) by various combinations of cold working and/or heat treatment. The temper designations for aluminum are not applicable to steel. Low carbon (no more than 0.30 percent carbon) and highly alloyed austenitic stainless steels are strengthened by cold work. Other steels are strengthened by one of three types of heat treatment: martensitic hardening, age hardening, and austempering. The maximum hardness of carbon and alloy steels depends on the amount of alloying elements, particularly carbon.

Although steels that are annealed (heated above the transformation temperature, held at this temperature for a specific period of time, then furnace cooled at a specific rate) generally are the softest and most ductile, a similar treatment called *normalizing* also is common. Normalizing consists of heating to well above the annealing temperature, holding at this temperature for a specific time, and then cooling in still air. This is done to refine the grain structure and put carbides into solution. Normalized steels are indicated by the symbol *N*. The conditions resulting from various combinations of work and heat treatment for steels are designated in various ways, depending on the category of steel under consideration. Some are classified by a description of the process, some by the strength level attained, and others by the heat treat temperature used.

Among the more common steels used for the manufacture of aircraft are those designated low alloy (carbon plus up to about 0.5 percent of various alloying elements such as manganese, silicon, nickel, chromium, molybdenum, vanadium, and boron). Steel alloys within the 4000 series (such as AISI 4130, 4140, 4340, 4330V, as well as D6AC) are in this family of steels and are chosen for their superior strength-to-weight ratios (that is, higher possible tensile strength than other alloys), through-hardening capability, and machinability. Through-hardening is the capability of maintaining the desirable level of at least 90 percent martensite (thus, more uniform mechanical properties) throughout the entire cross section of the part.

The hardening process varies from class to class of steel but, as an example, for the low-alloy steels it consists of a three-step operation. Step one consists of heating the part to above the transformation temperature and holding it at that temperature long enough to permit the alloying elements to form a solid solution, which is called austenite.

Step two consists of quenching the heated part in a quenching medium such as oil. This rapid cooling changes the structure from austenite to martensite, which represents the point of extreme hardness, but also extreme brittleness.

Step three brings about some softening, which is accomplished by tempering, heating the part to some intermediate temperature lower than the austenitizing temperature: the higher the temperature, the lower the final strength. Parts hardened to 200,000 psi or above that are machined or ground after final tempering and certain steels that are cold formed or cold straightened must be stress relieved. Stress relief is accomplished by heating the part to within a particular temperature range below the minimum tempering temperature and holding it at that temperature for a specified time to prevent possible crackling (in other words, to relieve unwanted stresses introduced during the processing).

Carbon steels, like stainless steels, come in many types. Note that the word is stainless rather than stainfree. Some of these steels do develop stains after environmental exposure, but not the degree of rust that is associated with carbon steels. Martensitic or semiaustenitic stainless steels are hardened by heat treatment and sometimes extensive cold working, have good strength and oxidation and corrosion resistance, and are designated as precipitation and transformation-hardening stainless steels.

Austenitic stainless steels, the basic grade of which (type 302) contains 18 percent chromium and 8 percent nickel (thus considered an 18–8 stainless steel) are not hardened by heat treatment, but can attain high strength levels after cold work. Varying the percentage of chromium and nickel and adding other alloying elements can provide special characteristics. These steels have excellent high temperature oxidation and corrosion resistance.

Heat-resistant alloys (such as A-286, various Inconels, and others) have greater than the 18 percent chromium and 8 percent nickel of the austenitic stainless steels and may have a base element other than iron. Properties are affected by changes in chemistry, heat treatment, and processing and generally exhibit adequate strength and oxidation resistance for use at elevated temperatures without special surface protection.

Titanium and Special-Purpose Alloys
Titanium is a relatively lightweight metal (about 62 percent heavier than aluminum, but only 58 percent as heavy as steel) exhibiting good corrosion

and oxidation resistance. It can be greatly strengthened by the addition of various alloying elements and, in some alloys, heat treatment. Two commonly used aircraft alloys, Ti-6Al-4V and Ti-6Al-6V-2Sn (the numbers representing the percentages of each alloying element such as aluminum), are used in either the annealed condition for maximum toughness or the solution treated and aged (STA) condition for maximum strength. The former alloy is more readily weldable than the latter, both requiring a thermal relief treatment after welding.

Many special-purpose alloys exist, including beryllium copper, manganese bronze, and MP35N. The latter, a multiphase alloy of cobalt, nickel, chromium, and molybdenum, can attain extremely high strengths with excellent corrosion and oxidation resistance. Its use is increasing for the manufacture of high-strength fasteners, superseding those of the more brittle H-11 alloy steel.

Further Processing—
Forming, Straightening, and Residual Stresses

The word processing has been used to describe the family of heat treatment, cold work, stress relief, and other techniques used to obtain a desired temper, hardness, or other characteristic of a particular metal alloy. It also is used to describe other steps in the cycle of manufacture of both metals and nonmetals. (See page 46.) Some attention to the processing of nonmetallics will be given in Chapter 5.

Additional processing of metals covers the field of forming or straightening, straightening assumed to be a sort of reverse forming (that is, the correction of bent or warped parts to a required straight configuration). Heat treat specifications often cover the aspects of straightening and may specify additional thermal treatment following the straightening to eliminate the stresses sustained during the mechanical part of the operation. The presence of such stresses, generally called residual stresses, is an undesired by-product of straightening and forming operations, as well as machining and grinding, and can cause undesirable aftereffects if not relieved or at least recognized. These stresses can lead to subsequent warping of a part (for example, if unevenly machined away) and will add to other stresses generated during the part's use.

The forming of metallic parts is controlled by the use of various specifications covering forming and straightening (hot or cold), peen forming, shot peening and air hammer peen forming, forming and aging, superplastic forming, and die quench forming. A quick reading of any of these specs that may be applicable to a particular misformed material is highly recommended, not only for the basic knowledge obtained, but for the addition of another building block to the storehouse of possible repairs for future use.

Hardness Testing—Steels

Hardness has been mentioned in relation to metal alloys, strengths, and tempers and some further explanation may be helpful. Original hardness testing was done by determining the relationship between the force required to push a hardened ball or calibrated probe into the surface of a piece of metal and the depth of the resulting depression. This remains a prime method and is particularly useful for the harder metals such as steel. Various instruments have been developed for this purpose. The more common of these instruments are the Rockwell and Brinel testers. Both specify the use of different-size balls and loadings for different metals, along with the equivalent material tensile strengths in psi, based on the hardness reading given on the instrument's dial. These manufacturers, as well as many using this equipment, have published charts showing the tensile strengths of various metals versus the hardness testing machine readings obtained. For example, a reading of 27 on the Rockwell C testing machine using a 150 kilogram load indicates the tensile strength at the surface of the metal being tested to be 125,000 psi or 125 ksi.

The accuracy of this reading could be verified by actually tensile testing the specimen being evaluated and recording the tensile stress at failure. In practice, many specimens would be tested and statistical variations taken into account. It must be realized that this strength is at the surface (when obtained from the hardness tester) and may be different elsewhere, especially if the surface of the part had been case hardened, had been unintentionally hardened by exposure to excessive heat (such as might exist after an unauthorized high-speed grinding operation), or was of such thickness that the material is softer at the center than at the surface. In evaluating surface hardness readings, the MRB engineer may be well advised to assume the temporary role of a doubting Thomas, especially if the reading is suspicious or substantially different from that anticipated from the test's circumstances.

Temper Testing—Aluminum Alloys, Electrical Conductivity

Aluminum alloys also are commonly tested for hardness and, as in the case for steels, must be of a minimum thickness and properly supported to avoid a false reading. The readings for aluminum generally are related to temper levels, as well as surface tensile stress (the temper being an indication of a range of tensile strengths, commonly a statistical minimum tensile stress). Temper levels of aluminum alloys also are measured by the use of an electrical conductivity test whereby an electronic tester measures the conductivity of nonmagnetic materials by generating an eddy current in the part being tested and measuring the change of electrical properties in the coil used to induce the eddy current. The absolute electrical conductivity obtained is presented in terms of the percent of the International Annealed Copper Standard* (% IACS) and varies from material to material and temper to temper. Tables are published in various sources including MIL-H-6088 and MIL-STD-1537, Electrical Conductivity Test for Verification of Heat Treatment of Aluminum Alloys, Eddy Current Method, and generally show allowable ranges of readings applicable to each alloy and temper.

Additional Metal Defects—Orange Peel or Alligator Skin, Sulphur Stringers, Decarburization, Hydrogen Embrittlement, Untempered Martensite

A few of the phenomena associated with metal parts, that are unwanted will now be discussed in detail. There may be many more. Among the least common is the appearance on the surface of a stretched or formed skin of a pebblelike surface called (and looking like) *orange peel*. This can occur when a material having an unusually coarse grain is formed or stretched beyond its elastic limit (in other words, beyond the point of return to its original size). This surface effect also may be referred to as an *alligator skin*. Consultation with metallurgists is recommended if this condition occurs.

The appearance of *sulphur stringers* on the surface of parts made of Type 416 corrosion-resisting steel was brought to the author's attention when voids were discovered on the chrome-plated surface of a part. The sulphur,

---

* A commercially pure copper rod one meter long with a 1 mm^2 uniform cross section having a conductivity of .017241 ohms at 20° C, considered to be 100 percent IACS.

added to the material from which the part was made to improve machinability and increase its resistance to galling, and originally in globular form, had been distended into a long stringerlike pattern during the hot rolling operations. Exposed at the surface after final machining, the nonmetallic sulphur did not permit the subsequent chrome plate to uniformly wet the surface of the part, resulting in a series of parallel voids in the chrome. The solution would have been to strip the plating, chemically etch away the sulphur, and then replate and grind to size, but the cost of rework made this an expensive option. Subsequent lots of parts were made of material from another supplier and checked for sulphur inclusions prior to chrome plating.

Decarburization is a term sometimes encountered in the material review field. It refers to the loss of carbon from the surface of ferrous (iron) alloys when they are heated during processing above 1300° F in a medium that reacts with the carbon at the surface. This can occur during hot working or heat treatments, such as normalizing or annealing, and can be controlled by performing these operations either within a protective atmosphere or on parts protected with a special coating. Another option is to machine away the decarburized surface (which may be up to .050 inch deep) prior to use of the part. Decarburization causes a substantial reduction in both the strength and the fatigue life of the part. In general, complete decarburization is not permitted although partial decarburization up to .003 inch deep is permitted. Carbon restoration is not allowed.

Hydrogen embrittlement is another unwanted phenomenon occasionally encountered and refers to the embrittlement of certain types of steels caused by the introduction of hydrogen into the part during improper processing. This embrittlement may lead to a premature failure under load and can result from improper processing during electrolytic cadmium plating, such as the use of too low an electrical current density, the failure to follow up the plating operation with an embrittlement relief baking operation (375° ± 25° F for twelve hours minimum within one hour after plating), or the use of acid pickling during preplate cleaning.

Another unwanted phenomenon is that generated when highly hardened steels (220,000 psi or greater tensile stress) are finished ground or finish machined after heat treatment in such a manner as to cause excessive heating at the part's surface. The result of such heating, often detected by the presence of discoloration at the surface, is an area of *untempered martensite*. It is extremely hard and brittle; thus susceptible to surface cracking. This heating can be caused by inadequately cooled grinding, the use of dull tools, incorrect tool speeds or feeds, or the unauthorized use of a heating torch or other local heat-generating device. The biggest problem is one of detection because unauthorized or involuntary actions on the part of operating personnel are not always voluntarily reported, especially if the operator is unaware of the consequences of such actions. As a result, there are various prohibitions against the use or generation of excessive heat when high heat-treat steels are ground or machined after heat treatment. Inspection personnel, in particular, are cautioned to look closely for signs of overheating.

What really happens when such parts are locally overheated is that they experience a local mini-heat-treatment cycle without the usually required tempering operation which serves to alleviate the effects of the extreme hardness. The surface of the part in the area of the local overheating will reach the austenitizing temperature and then, once the heat source is removed, the surrounding cooler areas of the part will act as a quenching medium, the temperature at the previously heated surface will plummet, resulting in formation of a local area of untempered martensite.

Once suspected, this condition may be verified by the use of a chemical test called a nital etch inspection which consists of immersing the part into (or swabbing on the surface) a series of solutions containing nitric acid and alcohol (NITAL) or water; hydrochloric acid and alcohol or water; sodium or ammonium hydroxide and water; or sodium bicarbonate and water. The surface area is then checked for discoloration, varying from white to light gray for indications of untempered martensite to dark gray or black as an indication of overtempering.

Experience is required to properly evaluate these discolorations and the only remedy, other than acceptance, which would seldom be recommended, is to carefully remove the damaged surface area and renital etch or reheat-treat the part. Resurfacing the part is difficult because of the extreme hardness and results in some loss of static strength, whereas reheat treatment may cause distortion. In all cases, the nital etch must be followed within three hours by a baking procedure for an additional three hours. Untempered martensite is a phenomenon to avoid.

Corrosion—Many Types

Many defects result from the action of corrosion, but the procedure that has allowed or caused the corrosion to occur is itself a defect. Corrosion is never desired, never intended (with rare exceptions like the use of a magnesium link that is intended to corrode away in seawater and permit the release of a lobster pot marker float after a desired period of immersion). In fact, billions of dollars are spent in attempts to prevent corrosion from occurring because the result of corrosion is weakening of the part.

The most common type of corrosion is oxidation, caused by the tendency of most metals to combine with the oxygen always present in air and water. The surface of the metal exposed to oxygen will turn into an oxide of the metal, generally indicated by discoloration or staining. The common word for iron oxide is rust. The blotchy or mottled chalky white deposits on the surface of many types of aluminum alloys are aluminum hydroxide. Certain metals will continue to corrode in the presence of oxygen, others will stabilize somewhat once the initial oxidation has taken place, and at least one type of maranging steel is intended to corrode in place to a rich brown patina, an architectural consideration. Additional exposure to salts, acids, alkalies, and various chemicals will often hasten the corrosion process, a side effect of acid rain (atmospheric moisture containing various compounds of sulphur and perhaps other chemicals) being a case in point. Most manufactured products are intended to be protected from the effects of corrosion by various means, but the protective barriers such as paints, platings, and so forth, are sometimes abraded or worn away during handling or later use.

The second most common type of corrosion is galvanic corrosion, which takes place when dissimilar metals touch one another in the presence of moisture (water). This sets up an electrolytic action with one metal (the more active in seawater) becoming the anode, the other metal (the less active in seawater, also called more noble) becoming the cathode, and the water becoming the electrolyte. The result is akin to a wet cell electric storage battery with the more active metal (the anode) slowly corroding away. If a piece of unprotected steel (anodic) is in contact with a piece of silver (highly cathodic) in a moist room, the steel will slowly disintegrate. A piece of zinc immersed in an electrolyte (such as a sulphuric acid solution) along with a piece of lead also will disintegrate, releasing a direct electrical current in the process when the lead and zinc are connected externally. This process will be reversed when an electric current is fed back into the cell, thus recharging it. For purposes of determining the relative galvanic compatibility of various metals, they have been tested and arranged in the order of decreasing activity in seawater, as well as other electrolytes. The most active (or at least noble) metal is at the top, or anodic, end of the chart while the least active metal is at the bottom (or cathodic) end of the chart. This listing is called the galvanic series.

The closer together in the series any two separate metals are, the less galvanic action will occur between them. If two widely separated metals can be effectively plated with closely spaced (on the chart) metallic coatings, the galvanic action will be comparably less.

Yet another type of corrosion is intergranular or intercrystalline corrosion, a galvanic type of action between the different metallic crystals within an alloy. The attack is selective at the grain boundary zone and may be further defined as interdendritic corrosion in cast structures, interfragmentary in a wrought unrecrystallized structure, and intergranular in a recrystallized structure. A variant of intergranular corrosion is exfoliation corrosion (sometimes entitled lamellar or layer corrosion), which occurs along grain boundaries, but usually along narrow paths parallel to the part's surface. It generally occurs within thin products that have been highly worked, thus having an elongated grain structure. Exfoliation corrosion appears as a lamination of alternate layers of relatively uncorroded metal and thicker layers of heavy blistering, these layers being thicker than the original and resulting in an overall thickness greater than before the exfoliation. The only action recommended is replacement of the part, at least in the damaged area.

Stress Corrosion

Two equally serious words that the structural MRB engineer should be familiar with are stress and corrosion, combined as *stress corrosion*. As the name suggests, this is an amalgamation of the effect of unwanted stress and corrosion. This can occur on certain types of aluminum parts where the stress is continuously applied to the part in a generally unrecognized manner and in an atmosphere conducive to the formation of corrosion.

Normal corrosion (if the term normal can be used; the author sometimes used the term *garden variety corrosion*) is a natural tendency of most materials to combine with atmospheric oxygen to form an oxide film on the part's surface. Different alloys of steel and aluminum vary in their resistance to the formation of surface oxides. On some alloys the oxidation process is continuous, whereas on others it tends to be somewhat self-limiting. Stress corrosion may occur, however, with no noticeable surface discoloration.

The first awareness of the presence of stress corrosion may be the discovery of an aluminum fitting or machined part lying in storage on a warehouse shelf and found to be already cracked, although never used. Other parts may be bolted or riveted in place on an assembly of parts and found to have cracked in position some time after installation, but before delivery to the customer and use in service. Yet other parts may be discovered to have stress corrosion cracks in service after only limited use. As in the case of untempered martensite in high-strength steel parts, an understanding of the causes of stress corrosion in certain high-strength, heat-treatable aluminum alloys is most necessary for the knowledgeable MRB engineer.

Simply stated, stress corrosion pertains to the tendency of certain (not all) high-strength, heat-treatable aluminum alloys, when subject to sustained (continuous) local tensile stresses applied in a particular orientation with the part's grain structure, to crack within a relatively short period of time when the part is exposed to a corrosive (moist) atmosphere. The atmosphere may be merely that of a normally humid day at the seashore; Death Valley humidity would not apply. The sustained stresses may be residual in nature (that is, the type that may exist when the part is unevenly machined from a larger billet) or possibly cold formed to a substantial curvature and not stress relieved. More likely, however, are sustained stresses caused by forcing an imperfectly fitting part into place during installation and holding it in place with assembly fasteners such as bolts, screws, or rivets. Manufacturing people frequently are cautioned to not apply excessive force to make a part fit, but instead to bring the condition to the attention of cognizant inspection or MRB personnel.

An alternative is to install a suitable correctly fitting filler (shim) within the gap caused by the imperfect part(s) if such a procedure has been previously authorized. The admonition should be shim to fit, not force to fit. If the part is forced into place with heavy pressure (such as could be applied by a C-clamp or the foot of a heavy man) and held that way by installing permanent fasteners, and if the atmosphere is other than bone dry, the part (if made of a type of aluminum alloy susceptible to stress corrosion) may subsequently crack in place.

An entire series of aluminum alloys and tempers not susceptible to stress corrosion has been developed to replace those that are, although they generally are not as strong.

The one remaining condition often necessary for a stress corrosion failure to take place is that the sustained stress be applied to the part in an orientation such that this stress is acting in the short-transverse grain direction, the weakest of the three grain directions of a rolled billet of material. Of the three grain directions present, the resistance to stress corrosion is the least when the load is applied parallel to the short-transverse grain direction and the greatest when applied parallel to the longitudinal grain direction, generally the direction in line with the direction of rolling when the aluminum billet from which the part was made was originally fabricated.

Various tests have been developed to determine the resistance of different materials to stress corrosion (reference ASTM G47). Many require a test specimen to be clamped in such a way as to sustain a constant tensile stress and then be alternately immersed in and removed from a liquid salt solution, noting the time to failure. After a series of such tests, the maximum tension stress that the particular material can sustain without a stress corrosion failure (sometimes called the threshold stress) can be determined and published in manuals such as MIL-HDBK-5, as well as various manufacturers' data sheets.

Chemical Milling Described

The technology of chemical milling is one that the MRB engineer should have some familiarity with since *chem-mill* defects also are presented to the MRB for consideration and disposition. Prior to the introduction of chem-milling sometime in the 1960s, the reinforcement of aluminum alloy (or other metallic) skins was accomplished by the addition of riveted-on reinforcements called doublers. As many as a half dozen or more doublers would be added to a structural member to allow for (or supply a resistance to) increases in the loadings expected to be applied to the structure, and a large skin assembly might be made up of fifty or sixty separate parts with a thousand rivets necessary to hold them in place and allow them to absorb a proportionate share of the load. With the advent of chem-milling, the fifty or sixty separate parts might be reduced to three or four.

Chem-milling is basically the chemical removal or etching away of metal from the face of a part when immersed in a suitable liquid caustic or alkaline etchant. In the aircraft industry it is done with steel and titanium, as well as aluminum, and can be precisely controlled both as to the depth of parent metal removed and the surface location from which the removal is to take place. Since the removal of aluminum from any surface exposed to the etchant is approximately .001 inch per minute of immersion time the removal of .030 inch from the face of an aluminum skin can be accomplished with a thirty-minute dip in the tank. Of course, this is subject to careful control, since the effectiveness of an etchant will vary with increasing use and contamination. Other liquid solutions are used both before and after the actual etch cycle for cleaning, rinsing, and so on, but the overall cycle including forced-air drying is not lengthy.

The area to be specifically chem-milled (seldom are all areas on both faces as well as the edges of a particular part intended to be etched) is controlled as follows. The part is dipped in or otherwise coated with a rubber type liquid maskant which is allowed to harden to a thin filmlike coating. Then that particular area to be chem-milled has the maskant carefully cut away with a special knife to permit later exposure to the etchant.

The precise location of the maskant removal is controlled by the use of a template, configured to guide the knife along the same curved or straight line as the line to be milled. This scribe line actually is set forward a specified amount from the final desired location of the chem-mill line since the etchant will remove the parent metal sideways under the mask as well as downward. In this manner, all sorts of highly curved or geometric patterns can be created, including those as complicated as an individual's name in script. On high-volume production runs, substantially more refined methods for maskant removal than the hand-knife technique can be employed.

An equally important capacity for the chem-mill process is that of step milling, whereby varying final depths of milling can be accomplished by repeating the cycle additional times on previously milled surfaces. Generally, the deepest pattern is milled first, with added areas (plus the previously milled area) etched during subsequent cycles. Prior to each successive cycle an enlarged pattern of maskant is removed until the last cycle removes metal from the least deep area in addition to all the previously milled areas. If properly done, the final part will contain the required number of correctly positioned recesses, each of the required remaining thickness or, in some cases, depth of cut (in other words, metal removal).

Another benefit of the chem-mill process is the ability to manufacture many parts at the same time, of either the same or a different configuration, as long as the type of material and the depth of cut are the same during the particular cycle and the tank is large enough to accept the load.

Chem-Mill Defects

Defects in either the location of the chem-mill line or the remaining thickness of material are among the common types brought to the attention of the MRB engineer and must be evaluated on the basis of the effect on the structural integrity of the part, both as to the basic reduction in strength and the effect of the missing material on nearby holes or free edges from the standpoint of the probable reduction in fatigue life. Certain other defects are unique to the chem-mill etching process itself and many are addressed within standard repair manuals. Among these are blowouts, pitmarks, and pockmarks.

Blowouts are unwanted areas of metal removal caused by the local lifting up of the rubber maskant along an imperfectly cut edge of the maskant and not detected or repaired prior to the submersion of the part into the etch tank. This usually results from the use of a dull knife or a lack of pressure on the blade during the cutting action, with the effect being a local lift-up of the maskant from the part's surface. This allows the etchant to act outward from the desired scribe line and remove an area of metal looking like a recessed half moon. The location and degree of undercut are important in determining whether or not the condition is acceptable. About the only thing that can be said in favor of acceptance is that the undercut will not be deeper than the adjoining milled area and the radius will be the same since the etching action proceeds about equally in all directions.

Pitmarks are similar to blowouts, but emanate from isolated holes in the maskant rather than from the edge of an otherwise sound chem-mill undercut. As such, they take on the shape of the maskant cutout, but retain the same depth of cut as for the dip cycle (or cycles) when they occurred. In extreme cases, the pitmark (if undiscovered at the beginning of a multiple dip chem-mill cycle) will end up as a hole completely through the part. Pitmarks often are caused by pinholes in the maskant, by undetected gouges in the surface of the part prior to the regular chem-milling, or by improper remasking of thickness measurement windows (called micrometer holes) cut into the maskant during the chem-milling operations to permit micrometer inspection of the part's remaining thickness. Prior tool marks or gouges will not etch any faster than the parent surface of the part being milled, but the final local recess will be proportionately deeper than the surrounding surface and wider than the original recess in all directions.

Pockmarks generally appear as long lines or chains of closely spaced but small individually circular depressions on the chem-milled surface of a part containing either impurities or the segregation of one or more of the alloying elements within the material.

These elements etch at a faster rate than the surrounding surface of the part and the alignment of the depressions usually is in the direction of rolling during the manufacture of the billet or sheet from which the part was made. Hence, the parallel lines of depressions. This condition is not to be confused with those causing pitmarks and usually can be verified by noting a change in the pattern of pocking as more and more metal is etched away from a sample review specimen. Pocks not originally exposed may become apparent as deeper etching takes place and those already exposed will deepen and widen, but only at the same rate as surrounding geographical features, assuming that all of the originating material has leached away. Pockmarks are not as common as pitmarks, but patterns of pits may be considered suspicious.

Oilcans

An *oilcan* is the term used to describe the condition of a skin panel when it can be pushed inward with only modest pressure and then pops outward again once the pressure is removed. The term, of course, is taken from the same effect when the bottom of an oilcan is flexed inward with the thumb to deliver a drop or squirt of oil.

The term oilcan also is applicable when the panel pops or clicks inward and stays that way, but will pop outward again when a similar light pressure is applied in the opposite direction. Much confusion exists between a true oilcan condition and normal panel deflection. Certainly the evidence of an audible light snap, pop, or click would suggest a true oilcan, whereas panel movement without noise would not. One of the problems with both oilcans and normal panel deflections is that they tend to vary with the surrounding support given the panel and the stage of manufacturing at which they are looked for or detected. In general, they are considered undesirable inasmuch as excessive repetition of the oilcan by forces applied at random or through long use may result in eventual cracking of the panel and ultimate failure. In the case of oilcans on aircraft, they may appear early in production and then disappear as assembly progresses only to reappear when the plane is resting on its wheels with no fuel in its tanks, or vice versa. The addition of repair stiffeners generally is accomplished at the point of initial discovery.

Tubing Defects

Defects on tubing are not uncommon although, when they occur on short or easily replaceable lengths, the damaged tubing may be discarded and replaced rather than repaired. Repairs generally are requested when the defects occur after installation on complicated assemblies where replacement is totally impractical or where substitute parts are not available within the time span allowed.

A distinction must be made between structural tubing (that is, a primary load carrying element like a strut or column) and tubing designed to carry fluids or gases (which may vary from highly pressurized hydraulic lines to mere drain lines experiencing little or no pressure). As in all MRB engineering work, one must know precisely what is involved. Another distinction must be the type of material and whether or not it is capable of local repair by welding (in the case of metal tubing) or by the use of adhesives (in the case of nonmetallic tubing).

A further consideration is the effect of a weld repair on the tubing itself and whether or not the locally annealed (hence softened) area next to the weld is acceptable. The most common tubing defects that cannot be repaired without engineering authorization are dents which can be caused by slipups during installation, dropped tools, and so forth. Some dents may be allowed by the applicable tubing fabrication or installation specification, but those beyond the stated limits must be dispositioned by the MRB, and the configuration of the dent (smooth versus sharp) must be taken into account. Punctures, often associated with dents, are a more severe type of defect as are holes which might be caused by a nearby drilling operation. Dents might be accepted without repair, but not punctures and holes in fluid-carrying tubing.

5 Nonconformance Types and Descriptions—Fiberglass, Composites, Honeycomb, Nonmetallics

Fiberglass Laminate Construction and Fabrication Techniques

Defects (nonconformances) in parts made of fiberglass tend to be generic in nature in that perfect tools seems to be in short supply and errors in workmanship occur (at least within the aircraft industry where small production runs seem to be the norm).

The term fiberglass in the context of this publication actually refers to a fiberglass laminate, a built-up part consisting of alternate layers of glass cloth and an adhesive, or more correctly a resin, subjected to adequate pressure and heat to cure (harden) the resin and form a continuous solid mass. The glass cloth itself may be used to make products such as window curtains and the fibers are widely used as one type of insulation.

An early form of structural laminate was plywood, containing alternate layers of thin wood sheet and adhesive. Later development of this technology led to the manufacture of curved panels of plywood, particularly in the marine field. Fiberglass laminates have the advantage of relative impermeability to moisture and since they are made up of very thin sheets (commonly .010-inch thick) of glass cloth they can be molded into relatively complex shapes and varied in overall thickness as required by the designer. There are many types of cloth and resin systems in use, but the types of defects are fairly constant.

As for other technologies, a brief description of the processing required is helpful in understanding the causes of these defects. The wet layup consists of alternate plies of glass cloth and a suitable resin applied by brush or spray or sometimes in the form of a thin sheet of adhesive film which must be carefully stored at low temperatures when not being used. This sandwich is built up layer by layer against a mold representing the shape of the surface of the final desired part; it is called a *bondform*.

In the case of a flat part, the bondform may be the flat surface of a special mechanical or hydraulic press called a platen press. Prior to placing the first ply of fiberglass cloth against the bondform, the form must be coated with a release agent or parting compound to prevent later sticking when the cured laminate is to be lifted off. The first ply of cloth is then laid against the surface of the bondform (sometimes preceded by a special peel ply meant to be removed later). This and all other plies must be rough trimmed to size and where necessary slit to allow for better fit around tight curves, following many of the techniques used in the garment industry. The inner surface is then coated with liquid resin, which is thoroughly wiped in between the strands of glass until the surface is fully *wetted*. In this manner, alternate plies of fiberglass cloth and coatings of resin are applied until the correct number of plies have been placed in position; more in the heavily reinforced locations (such as edge attachment areas) and less in many inside areas.

When the layup is complete, pressure is applied to the inner surface by one of several techniques. When the surface is flat and the final cured thickness of the laminate is to be constant, the movable upper plate of the platen press is lowered against the layup and pressure maintained. When the laminate is to be contoured the pressure is applied by the use of an additional rigid moldform mated to the bottom moldform, the two moldforms together are called matched metal molds.

In other cases, this pressure is applied by the use of a flexible plastic or rubberized bag against which an external source of fluid pressure is applied. This pressure is sometimes preceded by the introduction of a vacuum between the bag and the layup to permit a preliminary check of the process before rolling the assembly into the oven or autoclave for the final heat/pressure part of the cycle. Once the pressure is in place and all the excess resin squeezed out, heat is applied and held for a specified period of time sufficient to completely cure (harden) the resin. After cure and cool down, the assembled tooling devices are removed, the part lifted out, and final mechanical trimming and cleanup accomplished.

A less messy technique makes use of fiberglass cloth preimpregnated with resin (called prepreg) in place of the dry cloth and added liquid resin. Prepreg actually is a ply or layer of fiberglass (or composite) cloth or fabric coated with a resin that has attained what is commonly called its B-stage. The B-stage is an intermediate stage toward the final cured stage of a resin and enables it to be handled and processed before the required cure cycle, when it will attain its final physical properties. This system has both advantages and disadvantages and the choice would be up to not only the designer, but the tooling and manufacturing people since the final product is so dependent on control of the process and the tool expenses anticipated.

Fiberglass Laminate Defects Due to Improper Curing and Layup

The following explanation of defect types is also generally applicable to composites and honeycomb panels since many of the fabrication techniques are similar, requiring the controlled application of pressure and heat within a specified time span. The overall time period from the initial application of pressure or heat to the completion of cool down is called the *cure cycle*. Within this cycle the pressure and temperature must be carefully controlled between narrowly defined limits. Normally the range of increasing temperatures up to that temperature called the cure temperature, the rate of warmup called the heatup rate, and the time to be spent at the cure temperature are all specified, as may be the cool-down rate and the maximum temperature for removal of the part from the bondform. Any variation from these parameters, depending on the magnitude, may have adverse effects on the configuration and/or structural integrity of the finished part.

The dispositioning MRB engineer must determine first the magnitude of these defects and then judge whether or not they are acceptable. This normally requires close contact with the processing engineers and may require sample panels to be tested to failure; panels either made from scratch or one or more representative panels from the lot (group) of parts similarly misprocessed. For a lot of four similar or identical parts, perhaps subjected to a cure temperature 20° below the minimum acceptable cure temperature, the engineer would have one tested to save the remaining three, if the test could be anticipated to truly represent the effects of the low cure temperature, and the physical or other results of the test were within the acceptable range for the other three parts. Obviously, if there was only one part in the lot, the test, if it destroyed that part, would not save any other parts, but might still be of value to help build a data bank for similar defects that might occur in the future.

On an extensive long-term program, the number of different test results generated in an attempt to save various misprocessed parts may be substantial, and worth every penny, since scrap and replace seldom is the least expensive way to go, especially for large, complicated assemblies.

It should be understood that seldom is 100 percent of the specification strength of a part required for the part itself to still be acceptable. For example, the specification may require that the shear strength of a panel be 2500 psi, but the design strength required may be only 1900 psi. If the misprocessed panel fails at 2200 psi, the remaining panels may be acceptable if the test panel result is considered representative of the lot.

In a similar manner, other cure cycle defects (such as defective resin content) may be evaluated. Defective resin content results from the application of too much or too little resin during the layup operation, inadequate distribution of resin due to improper placement and wiping, or excessive localized bond pressure tending to locally squeeze out such resin. *Porosity* is related to insufficient resin content, but may be separately described. In reality, a correct resin content would result in a laminate without excessive porosity; porosity being the visual evidence of inadequate resin on the surface of the laminate. Insufficient resin inside the laminate may be the cause of, or result from, other defects. Areas of excessive resin are called resin-rich areas, whereas areas of insufficient resin are called resin-starved areas. Both areas may occur on the same part, often adjacent to one another.

The following types of fiberglass laminate defects often are brought about by improper fits between the individual elements within the laminate, or improperly configured tools or associated pressure application devices. The ideal laminate would be one with the correct number of individual plies of fiberglass cloth containing the precisely required volume and distribution of resin and squeezed together with the necessary pressure throughout the entire surface area in the presence of sufficient heat for the necessary period of time to produce a uniformly smooth solid laminate free of all bubbles, blisters, porosity, voids, bridging, and delamination. See MIL-HDBK-17B for the properties of fiberglass laminates.

If the application of pressure is to be brought about by the use of matched metal dies (male and female), the accuracy of these dies is extremely important. Often they are made up using computer-controlled machining techniques with the anticipated final thickness of the cured layup fed into the equations. What is tricky is the effect of heat on the male and female bond-forms, and substantial rework, adjustment, and so forth, might be necessary to account for the effects of heat warpage.

The cost of this process, along with the cost of the original matched metal molds, would be hard to justify for a low production rate program. For such programs, a more common technique is to manufacture a single bondform, either male or female depending on the complexity and amount of curvature, and then apply the necessary bonding pressure through the use of a flexible high-temperature-resistant plastic bag placed on the exposed surface of the completed but uncured layup. The edges of this bag are sealed all around the periphery with putty, tape, or other means and then the air between the bag and the laminate is evacuated with a vacuum pump.

Other plies of material, such as bleeder cloth (to adsorb excess resin squeeze out), also are employed in this process, but the effects of their use are beyond the scope of this discussion. Pulling a vacuum results in the atmospheric (air) pressure squeezing or pressing the bag against the surface of the laminate, compressing the laminate against the surface of the bond-form, squeezing out all excess resin (hopefully) and holding the laminate in this position through the entire cure cycle of heatup, cure time, and cool down.

In certain instances, and in most all cases where the fiberglass laminate is part of a more extensive assembly using honeycomb core (which is known as a honeycomb bondment or a honeycomb panel), additional pressure beyond the atmospheric pressure of around 14.7 psi is required. This usually is provided by use of an autoclave. An autoclave is basically a pressure cooker; a vessel capable of providing both heat and pressure to items placed within it. For industrial use, an autoclave may be large enough for a railroad car to enter. A fiberglass layup or honeycomb bondment placed inside an autoclave not only can be heated, but also pressurized to much higher than atmospheric pressure, since external pumps can pressurize the air or other gas inside the clave.

It is the uniform application of the bonding pressure and the proper placement of materials within the laminate that provides a sound final product. Improper pressure, mechanical hangups, insufficient plies of cloth or quantities of resin, poorly matched bondforms, improper trimming or placement of the plies (some of which may have been previously separately cured and thus are more rigid), and other problems can cause the following defects.

Bubbles or *blisters* are both terms used to describe lumps or local protrusions from the surface of a cured laminate. The implication is that the outermost ply itself is continuous (unbroken) but that somewhere inside the laminate one of the plies is separated from an adjoining ply resulting in another defect called a void. If the plies above and below the void are far enough apart from each other, the result is a local bubble or blister. Conceivably, the surface of a laminate could bulge outward without an underlying void, but this would be better defined as a lump or protrusion and all the plies in the laminate would be similarly displaced. Any defect described as a bubble or blister should be evaluated as both a bubble/blister and an underlying void. (See Figure 5.1.)

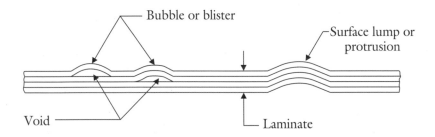

Figure 5.1.   Laminate defects.

A *void* is a local, often extremely thin, separation between the faces of adjoining plies of material not properly bonded together with the required resin or adhesive. It may be small or extremely large in area, but in any case no shearing load (parallel to the face of the parts) can be transferred between the plies, nor can any tension load or load required to keep the plies from buckling away from one another. Thus, it represents a structural discontinuity which may or may not be acceptable, depending on the size, location and structural need. Criteria for acceptability may exist, but in most cases the recommended repair action is to inject a suitable adhesive or resin into the cavity in an attempt to restore the strength and load transfer capabilities. Voids close to the surface sometimes are visible as lighter-toned, cloudy or milky-appearing blotches. More often they are detected as a result of specific inspection actions intended to ensure that the laminate is void free. The simplest of these inspections is the tap test where the inspector taps a special thin-handled hammer against the surface, moving it along the surface to detect any difference in sound.

The sound from a void-free area, depending on its thickness, will be relatively solid. Adjacent void areas will sound tinny or hollow, especially if the voids are near the surface. A fifty cent coin lightly held between the thumb and forefinger also makes a hand tap-test tool. The MRB engineer assigned to fiberglass, composite or honeycomb fabrication, or final installation areas should become familiar with this technique because an inspector may not always be available for diagnostic efforts requested by the engineer. Other more sophisticated techniques also are used for void detection using combinations of ultrasonics and electronics commonly called sonic tests. In all cases, the inspector's task is to locate the void areas and map out their locations precisely on the description of nonconformance paperwork, as well as on the part itself.

Bridging is a phenomenon resulting from the failure of a laminate to nest tightly against the bondform during the cure cycle and in an area or pocket of extreme curvature. It may result in a void, a delamination, a flat spot, or a bump in the cured laminate's surface. The laminate actually bridges across the curved area of the bondform, much as a wooden plank lying on the ground would bridge across a shallow hole in the ground. This condition is caused by insufficient pressure against the laminate during the early stages of the cure cycle, incorrect placement or trimming of the plies of cloth in the area of the sharp curve or corner of the bondform, or a hangup of the plies of the laminate near the bondform depression such that those plies are prevented from sliding completely into the recess. Special attention to the fit of the individual plies of the laminate may be required in these areas and special application of pressure may be necessary.

Delamination is the term used to more specifically describe a void between individual plies of a fiberglass laminate. It may occur only between two adjoining plies or between more than two adjoining plies within the total packup of the laminate.

Sometimes a delamination (delam for short) will either be revealed when excess material from along the edge of a laminate is trimmed away or it may be caused by the trimming operation itself. Generally, the strength of laminates is the least in the pull apart direction, (90 degrees to the part's plane) and the use of either dull tools or the lack of suitable clamping of the part in the area of the cutting operation often is the cause of an induced delamination. (See Figure 5.2.)

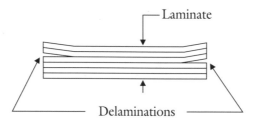

Figure 5.2.   Laminate delaminations.

Advanced Composite Description, Moisture Effects, and Defects

A review of the preceding information on fiberglass defects is highly advised before reading the following coverage on honeycomb and advanced composites. The term *advanced composites* (sometimes referred to as fiber-reinforced plastics) applies to laminates composed of fibrous materials embedded in a resin (such as epoxy, phenolic, or polyimide) matrix with the plies oriented in varying directions to provide specific strength and stiffness. These laminates represent the next generation beyond fiberglass laminates in that they employ much stiffer fibers of graphite, boron, Kevlar 49A (an aramid DuPont registered fiber), S2 glass, E glass, and others, thus warranting the term advanced.

Equally important is that the orientation of the fibers, generally applied in the form of tapes or fabrics, is particularly specified. Directions of 0°, 45°, 90°, and 135° are common rather than the random orientation of the fibers of many fiberglass laminates. A structural advanced composite such as used on a well-known forward-swept-wing research aircraft may have from three or four to over one hundred plies of tape in various directions and locations. A substantial body of literature has evolved for this technology, much of it proprietary, as individual companies have developed methods for design and analysis. One of the most helpful for the MRB engineer, however, is the Advanced Composite Repair Guide prepared by the Northrup Corporation for the United States Air Force Wright Aeronautical Laboratories at the Wright-Patterson Air Force Base in Ohio. Distribution, at least through 1982, was limited to U.S. government agencies and designated recipients.

Since advanced composites are of laminated construction, they are subject to many of the same types of defects as fiberglass laminates. Common to both are voids, porosity, and delaminations, as well as mechanical types of defects such as cracks, inclusions, wrinkles, and so forth.

Of particular note should be the effects of both moisture and heat on the strength of these materials at both room temperature (around 68° F) and at elevated temperatures. A material's so-called hot–wet properties often show a substantial degradation in strength from those at room temperature in a dry environment, and design parameters often specify acceptance levels of wet, dry, and various temperatures. Thus, moisture content is of great importance, not only insofar as it affects the strength of the laminate, but for the effect on initial fabrication and on subsequent repair operations.

A prerepair dryout or bake cycle is recommended because adsorbed moisture (these materials seem to have a natural affinity for moisture) can affect the integrity of repairs in at least three ways. In a honeycomb panel with trapped moisture, water will expand when exposed to the heat of the cure cycle (often above the boiling point of water) and the resulting internal pressure can either blow apart the nodes of the honeycomb cell walls (this is called *node blow*), or pop the skin or facing laminate away from the honeycomb core cell wall ends, causing a skin-to-core void. The possibility of this occurring is increased if the adhesives have already been weakened by the presence of such moisture and the cure cycle heat.

Another effect of this unwanted moisture is the tendency for moisture-caused porosity to develop in the bond line, thus weakening the joint. A corollary result may be the development of bubbles or blisters in the laminate which will cause a commensurate reduction in strength. The MRB engineer must be aware of the propensity for these types of defects in the absence of suitable moisture reduction controls.

A class of defects common to advanced composites are those related to the fibers themselves (that is, broken fibers, loose fibers, frayed fiber ends, and missing fibers, as well as delamination between piles), all at laminate surfaces or edges that have been mechanically cut away. These may be at a trimmed edge of the laminate or at the edge of a hole and in most cases are caused by an inadequate cutting or shearing action by the cutting tool. The tool may have dulled, the speed or feed of the tool may have been incorrect for the task at hand, or the thrust of the cutting tool may not have been backed up by suitable support against the backside of the part being cut.

Since backside support is costly and slows down production, it may not be provided for in all cases. Any internal hangup by the cutting tool may cause one or more of the plies to separate from their adjoining plies, resulting in a delamination and/or damaged fibers.

Honeycomb Bondment Description and Composition

Fiberglass laminates, advanced composite laminates and sheet metal skins often are adhesive bonded or, in the case of high-temperature-resistant metals, brazed onto a core made up of honeycomb material. The honeycomb core may be metallic or nonmetallic and the assembly, consisting of exterior skin, core, and interior skin, is called a honeycomb bondment, a honeycomb panel, or a sandwich structure. The word exterior (skin) generally only applies to a honeycomb panel when that particular skin is an air passage skin on the vehicle in which the panel is to be installed. Sometimes the skins are identified as upper skin (upward as installed on the vehicle), lower skin, inboard skin, and so forth, or bondform side skin or bag side skin depending on whether it is placed against the honeycomb bondform or on top of the core, just below the pressure bag. The term skin refers not only to a one-piece metal skin, but to a laminate used in the same manner, whether it is precured before being bonded to the core or is cured during the same heat/pressure bond cycle.

The prime advantage of a honeycomb panel is the weight saving accomplished by use of the relatively lightweight core and the ability of the core (formed into a honeycomblike configuration) to prevent the adjoining skins from buckling when properly bonded to the core. The core and skins also can be formed into complex curves and varied locally to match the load requirements of the panel. The spacing between skins also provides superior stiffness for lesser skin thickness since the moment of inertia (more on this later) varies as the square of this spacing.

The most predominant materials used for honeycomb panels on aircraft are aluminum alloys for both core and skins, although nonmetallics are increasingly coming into use. The skins frequently are chem-milled, with the thinner areas in the center of the panel and the thicker areas around the periphery where fastener attachment takes place and the skins are used to reinforce adjoining structural members. For air passage skins, the chem-milled steps usually are on the skin's inside surface, requiring the mating surfaces of the core to be similarly recessed—a neat problem for mechanical coordination.

The cores are made up of thin strips of aluminum alloy or other material formed into a zigzag, toothlike configuration and adhesive bonded to adjoining parallel strips at the mating faces which are called *nodes*. Each pair of adjoining strips, once bonded together, make up a honeycomblike pattern, usually six-sided in the shape of a hexagon and forming a lightweight (the strips are only a few thousandths inch thick), but stiff, matrix. The direction, parallel to these strips or ribbons, is called the *ribbon direction*; the direction 90° to the ribbon direction is called the *transverse direction*. The strength of the core is greater in the ribbon direction than in the transverse direction. This is partly because no load transfer across adhesive lines is necessary in the ribbon direction. The strength overall is affected by the thickness of the individual pieces of foil, the size of the individual cells (as measured by the diameter of the circle that could just be positioned within the hexagonal outline), and the type of material from which the ribbons or cell walls are made. It also should be noted that both the cell wall corner bends along each ribbon and the attachment of the ribbons at each node support the cell walls against buckling.

Core material is identified not only by the type of foil used but by the size of its cells. It is described by its weight per cubic foot. Thus, six-pound core would be twice as heavy as three-pound core, although not necessarily twice as strong. It may be further described as corrosion resistant if the foil has been treated with a corrosion-resisting compound either before or after the fabrication of the core.

As in the manufacture of fiberglass or composite laminates, the manufacture of a honeycomb panel requires a bondform and peripheral supporting members to keep the edges of the panel from sliding sideways under the bonding pressures, and to keep the bondment's edge-supporting members in the correct position. The closure of most honeycomb panels is accomplished by the use of sheet metal, machined metal, or fiberglass edge members in the shape of a channel or zee. One flange of the closure or edge member is bonded to the bondform-side skin inner surface, the other to the upper surface of the bag-side skin, and the web to the honeycomb core.

Individual pieces of core of the same or different density also are bonded to each other. Sometimes a special core having a cell shape more like a diamond than a hexagon is used where extreme flexing or curvature is required, and sometimes individual sections or pieces of core are bonded one on top of another, as well as the more common side-by-side arrangement. In all cases, every joint or splice is adhesive-bonded, each to its adjoining member. One source for honeycomb design data is MIL-HDBK-23, Analysis and Design for Sandwich Construction.

Honeycomb Bondment Defects Due to Improper Layup

When adhesive has been inadvertently left out or not been properly applied, when mating parts slip apart during the bond cycle, or were not properly fitted in the first place, a void will be developed; at that part of the joint, no load can be transferred. Discovery of these after-bond voids can be a tenuous matter since the closed honeycomb panel hides them from view. For the sake of this discussion and since the most common honeycomb panel is somewhat sandwich shaped, with its length and width much greater than its thickness, the term vertical will be considered parallel to the thickness, as these panels are commonly laid up flat or horizontally rather than on edge. In this manner the force of gravity will hold the individual parts against the bondform as fabrication proceeds and until the bonding pressures are applied.

A v*oid* can be defined as an absence of required adhesive. A *gap* can be described as an unwanted separation beyond the maximum prescribed or allowed. Much confusion exists between the two terms. An excessive gap filled with adhesive may be described as neither, but must be reviewed by the MRB engineer because excessive gapping between parts, whether or not filled with adhesive, may have an adverse effect on the joint's load transfer capabilities. A gap only partially filled with adhesive may be described merely as a void. To be correct, both the words void and gap should be used together when either zero or partial fill exists within a gap.

A void between two pieces of core is called a core-to-core void, sometimes referred to as a C/C void.

The adhesive often used between two pieces of core is a special type called *syntactic foam* or *expandable adhesive*. It is applied in the form of a thick sheet or ribbon (thickness varies) and contains a foaming or blowing agent causing it to expand substantially as it is heated. This has the advantage of filling local gaps between the cut ends of adjoining core pieces, thus reducing the possibility of voids. These cut cell wall ends seldom line up with one another across the joint, so the use of syntactic foam is almost mandatory.

A similar type of void is one between the edge of a section of honeycomb core and the vertical web of its adjoining closure member; often referred to as a core-to-vertical leg or C/VL void.

These may occur as one long, continuous void, but more often occur as small local voids individually separated along the required bond line. When continuous they often are the result of core-to-edge member gaps.

Voids between the inner surface of the face skins and the adjoining honeycomb core cell ends are called skin-to-core or metal-to-core (M/C) voids when the skins are metallic.

The adhesive used between skins and core usually is in the form of a sheet or blanket, consisting of a think layer of adhesive film squeezed against both sides of (actually embedded within) an inert carrier cloth like an open weave piece of fiberglass. This blanket is supplied in the form of large rolls resembling carpeting and must be kept refrigerated until use, lest it start to harden or cure at room temperature even though it may require an actual curing temperature of many hundred degrees for the final bond. Skin-to-core voids, like the others, mostly result from improper fits (thus inadequate contact) between skin and core. The core may be locally undercut too deeply, or inadequate pressure against the sandwich in the localized area may be the problem. Occasionally, the surface of the skin and/or the core may have been dirty, thus preventing an adequate bond. This is difficult to detect because as long as the skin and core and the adjoining adhesive film are in physical contact with one another the (propensity for a) void cannot be detected by normal means. Long-term use may eventually reveal such a void that had escaped detection otherwise.

A similar type of skin-to-core void (note the temporary departure from the definition of a void as an area of missing adhesive) is one caused when one or both of the protective layers of film (attached to the outer faces of the adhesive at the time of manufacture to keep the adhesive from sticking to the adjacent material in the roll) have been left on when the adhesive was laid up. Both of these protective sheets must be removed before the final bond. The outer sheet, however, sometimes is inadvertently left in place, or a small piece may remain when the remainder is removed. The skin would then be laid in place and the bond cycle undertaken, with no indication of a void until much later, if at all. The only sure way to detect this condition would be to cut away a piece of the skin within this area and measure the force required to lift it off the core; in effect a pull test. Not only would the failing load be less than otherwise, but a careful observation of the test specimen would reveal the presence of the unwanted protective film, of a different color than the adhesive itself.

The use of the term *void* would be correct to describe the separation of adhesive from the part against which it was meant to adhere (the result is a gap), but this would be the end result of a substandard bond line. Sequentially, we would have the presence of a foreign material within the bond joint, a premature adhesive failure, and a detectable gap causing a void. This series of events, although implied, is seldom stated in full and it is often the task of the MRB engineer to mentally or physically reconstruct the past.

A final type of void is the metal-to-metal (M/M) or skin-to-solid adherend void, which is a void between the inner face of a honeycomb panel skin and a mating doubler, splice strip, skin, flange, block, or other solid member not of cellular or honeycomb construction.

The same adhesive film used between skin and core generally is used between solid adherends, but the continuous surfaces on both faces of these metal-to-metal-type joints may affect the quality of the bond line differently than a skin-to-core joint, where the cavities within each core cell offer a differing resistance to the bonding pressure.

Honeycomb Bondment Defects Due to Mechanical Force, Excess Internal Pressure, and Moisture Intrusion

The quality of honeycomb panels is very sensitive to processing variables such as temperature, pressure, fit-up, and time. Thus, variations beyond the cure cycle limits are the frequent cause of defects in the same manner as for fiberglass and composite laminates. The fabrication and frequent testing of process control panels and specimens is a necessity for the maintenance of high quality standards. Once completed, however, honeycomb panels are subject to the same mechanical damage as other parts and panels. Dents and depressions are all too common, especially within the thin skin areas of these panels. The author is convinced that some of these dents are caused by the elbows of people leaning against these panels. Underlying these dents and depressions may be matching areas of crushed core where the inward movement of the skin caused the vertical cell walls to buckle downward and sideways, either returning with the skin to the final dented position or forcing a crack in the bondline between the skin and core—a skin-to-core void.

A tap test may indicate a void or mistake an area of crushed core for a void. An X ray would reveal crushed core, but not a skin-to-core void. Ultrasonic testing would reveal a void but not a crushed core.

Two further types of defects worthy of description and consideration are blown core and moisture penetration. The resistance of the individual ribbons of honeycomb core to separation at each node from the adjoining ribbon at the same node is rather modest. This is especially so when the node adhesive is subject to the type of force least well resisted by adhesives in general; a peeling force like that when an individual tries to remove a Band-Aid in the least painful manner. Any pressure introduced inside the cells of honeycomb core in excess of the node resistance pressure will cause the mating pieces of foil to separate at the node, often in a zipperlike manner along an entire chain of in-line nodes. The result is a teardrop-shaped pattern of deformed cells called *blown core* or *node blow*. (See Figure 5.3.)

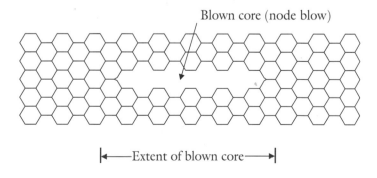

Blown core (node blow)

|◀——Extent of blown core——▶|

Figure 5.3.   Blown honeycomb core.

Sometimes associated with skin-to-core voids, which also can be caused by excessive internal pressure, blow core is most readily detected by X ray, which must be undertaken after the damage has occurred, not before. Excess pressure inside the core can occur if a mechanic uses a high-pressure air hose to clean away manufacturing debris or if a high-pressure autoclave bond cycle bag failure allows the same pressure to enter an opening into the core.

Similar damage can occur due to the presence of *moisture* inside a honeycomb panel, especially if this moisture has been exposed to heat above its boiling point as may occur during a honeycomb repair bond cycle or during high-speed flight. This seldom is a problem with a newly fabricated panel, but frequently can occur on a panel in service. Such damage also can occur when internal moisture in the form of trapped water is subject to freezing or alternate freezing and boiling. With proper sealing and no unwanted penetrations into the core of a honeycomb panel, moisture will have no way of entering, but service over time tends to generate dents, dings, and punctures and these may not always be adequately sealed at the time of occurrence. Water adsorption is insidious and studies with radioactive tracers have shown continuous migration of moisture over a long period of time.

In addition to the possibility of blown core and skin-to-core lift off, the very real effect of *corrosion* exists and often cannot be seen until either a structural failure or a pinhole in a metal skin is discovered. The author has witnessed extreme corrosion in aluminum honeycomb core with large areas of core reduced to power in the form of aluminum hydroxide, often accompanied by badly corroded inner skin surfaces. In cases like this, once discovered or at least suspected or anticipated, a major inspection and remanufacture effort is required. Unfortunately the cost of such an effort is substantial, requiring techniques such as radiography and ultrasonics. Fortunately, the increasing use of more corrosion resistant materials and better finishes, sealants, and sealing techniques in the newer designs reduces the propensity for the occurrence of these moisture-related problems.

Electrical Component Defects—Replacement, Splicing

Defects on electrical components frequently are handled on a remove and replace basis, especially since many, if not most, components are purchased parts and manufacturers' warranties may be invalidated if unauthorized repairs are undertaken. Replacement also is a requirement of some specifications, and requests for repair action may be generated only when removal and replacement is either impossible or totally impractical from a time/cost standpoint and (equally important) when a repair is both practical and suitable. Most requests for repair action, at least from the standpoint of aircraft production, are due to problems associated with the wiring between various components. Occasionally this wiring, especially on new production, may be too short, in which case a new length of wire may be spliced in to the component. In other cases, the insulation, shielding, or both may be damaged, or the circuitry on a printed circuit board may have been damaged. In these cases, an alternate circuit path is necessary, requiring the fabrication and installation of a jumper wire or cable. Therefore, the basic action is the provision for splicing in a replacement or alternate conduit. This often can be accomplished by reference to a standard repair, accompanied by any safeguards required by the MRB engineer for the particular application. (More sophisticated techniques in use are beyond the experience of the author.)

Fabric Component Defects—Replacement or Repair

Defects also occur on parts made of fabrics, parts secured to each other by age-old techniques (such as stitching), or by newer adhesive bonding techniques. Defects such as rips, tears, punctures, and cuts are not uncommon, but since fabric parts are not generally of high dollar value and by their nature are readily removable, repairs often are not called for unless a replacement part is not readily available. In this case, a standard repair may be utilized if the damage is within any defined standard repair limits. If not, the MRB engineer must design a repair for the mending or reinforcement of the damage utilizing, as is required for all repairs, his knowledge (either existing or newly developed) of all applicable splicing or joining techniques.

Transparencies—Defects and Acceptance Criteria

Transparencies—panels intended to provide for the transmission of light, a clear view of objects beyond, and protection from the elements—also are subject to various defects both during manufacture, handling, and subsequent use. These panels vary from small lenses used over instruments and lights, to large enclosures in the form of windshields and canopies. The transparencies usually are mounted in or supported by a structural framework and may, themselves, be a structural element carrying loads due to aerodynamic or other forces causing panel bending, shear, and hoop tension.

Defects on the smaller transparencies, unless acceptable by reference to an appropriate specification, usually result in the part's removal and replacement. Larger transparencies discovered to contain a defect may be so identified by inspection and submitted to the MRB for consideration, especially if the nature and magnitude are such that the personnel involved consider the unit to be either repairable or possibly acceptable for use as is by the MRB. Otherwise, removal and replacement is required, especially if the transparency is structural in nature.

Transparencies are subject primarily to surface contact damage causing pits, nicks, and scratches. Dents cannot be sustained because of the relative brittleness of the usual plastic or glasslike materials from which typical transparencies are made. A force large enough to cause a dent would generally crack or shatter the panel, rendering it unrepairable. Some transparencies are manufactured from a single ply of material such as acrylic. Others are in the form of a laminate like the typical automotive safety glass, but more sophisticated, containing three or more layers of material bonded together with transparent adhesives.

Simple panels are flat; the more complex ones may be sharply curved during the manufacturing process. When the panels must permit optical viewing, the effect of the defect on visual acuity must be taken into account and any defect that adversely reduces this visibility factor after a possible repair must be evaluated.

Occasionally, an internal defect within or adjacent to the adhesive or sealant line between plies will cause a visual impairment such as cloudiness or waviness. If this effect is beyond the acceptable minimum and is located within what is called the primary viewing area of the transparency, the panel must be replaced. The usual repair action for minor surface defects, the blend out and polish, also must be subject to the same visual test. The acceptance criteria are most stringent when the defect occurs within the primary viewing area, sometimes called optical zone 1. Less severe criteria generally exist in the secondary viewing and nonviewing areas.

Where visual acuity is a requirement, the effect of a nonrepairable defect or a defect after the accomplishment of a repair can be measured by various optical tests. A common test requires the viewing through the transparency of a screen containing a grid of closely spaced, 90° intersecting, parallel, narrow black lines painted against a white background. If distortion exists, the lines will exhibit a local interruption in their continuity, formerly straight lines appearing curved, formerly curved lines appearing wiggly or zigzag in outline.

Defects in Opaque Plastic Material

Defects also occur in parts made of opaque plastic material in either sheet form like ABS (acrylonitrile butadiene styrene), or in the form of cast or injection-molded material. Many of these parts are of low dollar value and would be replaced rather than repaired. The defects usually are mechanical in nature, such as mislocated holes, cracks, and so forth. Specific knowledge of the capabilities of these materials to bond to certain adhesives is required to effect the usual repair. They often are not amenable to mechanical fastening because of their brittleness.

Sources of Defect Listings and Definitions

Sources of defect listings and definitions are various manufacturers' standard repair manuals. The Military Structural Repair Manuals issued separately for each military aircraft in use contain definitions for many types of defects. The military also issues repair manuals for particular types of structures, such as honeycomb bondments. The MRB engineer working on a particular assignment should have available or obtain access to the repair manuals applicable to that particular job.

6 Written Descriptions of Nonconformances

Importance of the Write-Up—Need to Describe Both Observed Condition and Required Condition

One of the most important measures by which the quality or effectiveness of a material review system can be judged is the precision by which a defect is described. Ideally, the description of a defect or grouping of defects is so clear, complete, and correct that an experienced MRB engineer serving as an auditor or reviewer of the defect description document twenty years after it was first written can reconstruct the original circumstances with no questions or doubts in place. The only other documentation assumed to be available at the time of the review would be copies of any necessary engineering drawings or specifications referred to within the nonconformance write-up document. Anyone who has had to review volumes of old nonconformance documents years after the initiation dates will quickly come to understand the value of clear write-ups. The average collection of documents when studied religiously for more than several hours at a time leads directly to failing eyesight and increasing frustration. Perhaps everyone tasked with writing either the defect descriptions or the dispositions of nonconformances should be given the assignment to review old documents for a brief trial period, then submit a summary report.

The prime purpose of the nonconforming material document, be it an abbreviated or expanded format, is to clearly and completely describe the nature of the defect in terms of *actual* versus *should be*, often referred to as IS-S/B. It cannot be assumed that the reader already knows what the engineering requirements or S/B really are, although this may be the case if the reviewer is an experienced MRB engineer or quality engineer. The description of the S/B condition is generally that against which the inspection takes place. The description of the IS condition by nature is often more difficult because of the additional variables that may exist. The MRB or quality engineer who thinks she has seen it all hasn't been around long enough.

The description of a defect should relate to the original requirement as much as possible, but that is only possible if the defect is an enlargement, variation or modification of an existing geographical, physical, chemical, electrical, or other type of requirement.

The IS-S/B dichotomy is easy to delineate when the thickness of a particular tang on a machined fitting measures .109 inch whereas the engineering requirement is .125 ± .010 inch (.115 inch minimum), or the output voltage of a particular generator reads 22 when the minimum permitted by specification is 23. We can then easily define the defect in terms of a measurement versus the requirement, whether from the applicable engineering drawing or specification.

The requirement often is stated as per blueprint (B/P) or per specification (spec). There doesn't seem to be any consensus as to which comes first, the requirement or the actual, with the choice left to the writer. In all probability, the first written may be the first observed or measured. Should-be values are not always known with precision at the time of the initial inspection or discovery of an obvious defect. Many defects are discovered during a preliminary or unofficial inspection conducted by the mechanic entrusted with the manufacture of the part, or sometimes by supervisors.

The defects that are more difficult to describe are those that are outside the expected envelope of configurations or outputs, those that are not already defined by the requirements documents, be they drawings or specifications (which are listings of requirements). Thus, a bubble observed on the surface of a flat fiberglass laminate would be more difficult for a newcomer to the inspection department to measure and describe if no mention of bubbles was made on the B/P, than if Note 6 of the blueprint stated, "No surface bubbles larger than .50 inch permitted."

Write-Up Requires Description, Size, Quantity, and Location

The four requirements for proper definition of a defect are description (identification), size, quantity, and location, not necessarily in that order. The need for this information for each defect suggests that there should be a limit placed on the number of defects presented on any single document, or at least a requirement that all defects be related in some manner. From the standpoint of the reviewing engineers, the fewer defects per document, the better. From the standpoint of the initiators of nonconformance documents and those who count numbers of documents rather than numbers of defects, the more defects per document, the better. Practice varies; often depending on the number of defects known about at the particular point in time when the initiators realize that they had better start getting some answers.

The description of a defect must be such that everyone involved understands the defect. Common terms in use are the best, and perusal of governing specifications is a good starting point. Definitions often are provided in such specs, but when not available consultation with quality engineering, MRB engineering, or both, is recommended. Chapters 4 and 5 attempt to highlight some of the more common defect types.

Equally important to the description is the size of the defect since size often is a prime determinant of whether or not a particular defect is acceptable without repair. The usual measurements for the size of mechanical defects are length, width, and thickness or height. Length usually is the largest or longest measurement; width, the lesser of the two measurements in the same plane; and thickness, the dimension of the defect at approximately 90° or at a right angle to the plane described by the length and width. The term height usually describes the maximum rise of the defect from its base, whereas the depth of a defect usually would describe the unwanted extension of the defect (such as the depth of a scratch) into the body of the part that should not have experienced the defect. Verification is highly recommended. Determination of size generally is the responsibility of the inspection department, but a six-inch scale carried by the MRB engineer can be very useful as a rough backup.

It would seem that reporting the quantity of each (separate) type of defect is mandatory, but the author has seen nonconformance documents using only the terms "various" or "several," and occasionally "as marked on part." Needless to say, the documents were returned through channels to the initiator for clarification and proper definition. The author also has seen documents with six defects, but only five dispositions. Defect numbering is helpful to everyone and when mixed types of defects are included on the same document it is recommended that different numbering systems be used for each type of defect.

Thus, a particular document might list dents 1, 2, and 3 and voids A, B, C, and D. In some instances, the author has added an index to the document, listing the type of defect, the identification number of letter for each defect, the page number where the defect is described, and the page number where each defect is dispositioned. On a 50-page document, this certainly is worth the effort.

At least as important as the size of the defect is its location. Unfortunately, this often is the least adequately described and the most common reason for returning the nonconformance document to the initiator for more information. Many MRB instructions require that the location of a defect be defined by its physical coordinates in relation to the set of coordinates used to describe locations on the article (such as an airplane) on which the defect has occurred. Different articles use different coordinate systems with some commonality between boats and aircraft. The front to back locations generally are described by station numbers in inches, a position at or near the front of the vehicle being designated as station 0 and the extreme back end of a vehicle 100 feet long being described as station 1200 (that is 1200 inches aft of station 0). Vertical locations may be described by the use of waterlines, obviously descended from the nautical. Waterline 0 may be on a horizontal plane through the approximate centerline of the vehicle, front to back, or it may be a similar plane at the bottom or ground line or other reference point. Station numbers may be negative if the original design was extended forward, and waterline numbers may be plus if above waterline 0 or minus if below. Sidewise locations generally are identified by left-hand (LH) or right-hand (RH) buttlines with buttline 0 at the center of the vehicle from front to back (fore to aft). Left-hand would be as viewed from above or behind the vehicle and looking forward. A defect may thus be described as located at station 600 (600 inches aft of the front end), waterline 24 (24 inches above the horizontal, WL 0, reference plane) and buttline 42 RH (3 1/2 feet or 42 inches to the right of the vehicle centerline when viewed from behind or above). These are the three required coordinates, the X, Y, and Z coordinates of the defect. Most engineering drawings of groups of adjoined parts, called assembly drawings, and their related work orders to the manufacturing floor show the X, Y, and Z coordinates necessary to the design, as do the drawings showing adjoined assemblies and/or related equipment, and called installation drawings.

Location by Use of Coordinates Alone Often Unsatisfactory

Many drawings showing individual (detail) parts do not show the locating coordinates, or at the most show only one of the three coordinates for reference information alone. Additionally, the stations, waterlines, and buttlines are not painted, scribed, or otherwise marked on the parts or assemblies, so the only way of identifying them is with unique knowledge of which particular parts are supposed to be at which particular locations. A major fuselage (body) bulkhead (in an airplane like a wall between rooms) behind the pilots compartment may be at station 141, sometimes also called fuselage station (FS) 141. Another major bulkhead farther aft may be at FS 275, but there are no marks on the fuselage showing where FS 186 is, for example, so a listing of a nonconformance as located at FS 186 should be considered as advisory only. It is not uncommon for a floor inspector to mislocate the actual position of the defect by an inch or more, an accuracy totally unacceptable if the defect occurs at one particular hole out of a pattern of many holes and the MRB engineer must base the review on the loads applied at that particular hole, not the one to the left of it, or above it.

The use of coordinates should not be discouraged because it locates the field of play, but more specific locational descriptions must often be used, especially where the nonconformance is near or at a part of the vehicle where a distinct change in the shape, size, or configuration of the damaged member also occurs. For example, some parts will vary in thickness from, say .062 to .125 inch, along a line extending from the top to the bottom of the part. A gouge in the .125-inch thick portion of the part may be far less severe than if it had occurred within the .062-inch thick portion of the part, only one-half inch away.

Similarly, the occurrence of an enlarged diameter hole at the end of a part may be considerably more critical than the identical enlargement of the next hole farther away from the same end of the part.

Better Descriptions Utilize Sketches, Viewing Point, Marked-Up Drawing Copy, Photographs, Rubbings, Photocopies

To better define the location of a nonconformance it is highly recommended that a freehand sketch of the defect and the shape or outline of the area of the part within which the defect occurs be made and included along with the defect's description. The distance along the face of the part from a recognizable geographical feature (such as an edge) of the part to either the approximate centerline or to the edge of the defect also must be shown on the sketch. To really nail down the location, another dimension should also be included, perhaps the distance from the end of the part or from the centerline of a nearby row of holes to the center of the same defect. Ideally these locating dimensions should be at approximately right angles (90 degrees) to each other and the viewpoint from which the sketch was made should be written on the sketch's face (for example, view looking up, down, forward, aft, inboard (from outside looking inward), outboard, and so on). (See Figure 6.1.)

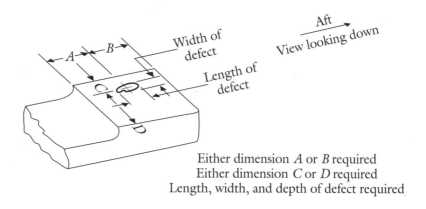

Either dimension *A* or *B* required
Either dimension *C* or *D* required
Length, width, and depth of defect required

Figure 6.1.   Defect size and location.

The view should be from where the observe is standing (or perhaps lying) in the most advantageous location to look directly at the nonconformance. A view from a location impossible to reach requires too much visualization to be effective. Don't hesitate to use directional words as just suggested. In fact, the sentence "center of dent four inches above lower end of part and two inches aft of centerline of most forward row of holes on part" is just as good as the placement of the same dimensions on the sketch.

` For those whose sketching talents are not equal to the task (the majority of those entrusted with the job of writing up the descriptions of nonconformances) an equally, perhaps more, effective method of presentation is to secure a copy of the engineering drawing or blueprint showing the area within which the nonconformance is located, and make a photocopy of that particular area of the drawing. Then mark on the copy the approximate location of the defect and describe its distance from existing geographical features as before. If the nonconformance is at a particular feature already shown (such as an oversized hole), it would be correct to circle or put some other mark across the view of the hole and add the words "this hole" or "hole shown circled."

If different holes were of different enlarged sizes, the holes would have to be identified as 1, 2, 3, 4, or A, B, C, D, and the actual size for each different hole would be listed on the document.

In some instances a photograph is the best way to document a nonconformance, especially if the costs of repair are expected to be borne by another company or the details of the nonconformance anticipated to be legally disputed. This is rare but, if in doubt, check with the quality or engineering members of the MRB.

Another effective way of documenting a nonconformance is with the use of a rubbing, especially if the defect is of a type to leave a distinctive mark on the sheet of paper used for the rubbing. Locations of holes, cracks, gouges, or any other nonconformances resulting in a depression or recess in an otherwise smooth surface are excellent candidates for a rubbing. Since the rubbing is a full-scale reproduction of the actual surface under consideration, the inclusion of dimensions is not as necessary, but a description of the surface location and directional data or arrows still should be added to the sheet of paper on which the rubbing was made.

Another technique that is very effective, especially when a visible defect occurs on a thin, relatively flat part, is to place the part on the viewing surface of a reproduction machine and make a photocopy. Try it on your hand first, then add the same directional data to the paper copy. Varying the exposure until a proper contrast results can provide an effective full-scale, reduced, or magnified picture of the defect and the surrounding geometry, but be sure to indicate the scale of the copy if not full scale. This technique can be especially useful for extra holes, mislocated holes, miscuts along the edges of parts, and so forth.

Underlying Defects—The Dashed Line Inquiry

An area of the defect description write-up that is sometimes missed by the initiator is the effect of the nonconformance, or even the existence of a separate nonconformance, on underlying parts or structure. Such parts may or may not follow the outline of the outermost part in a stackup (or packup) of parts and, unless they are subsequently removed from the packup for additional work, a defect on such a part may not be discovered. One example would be edge distance from a common hole that would be less on the underlying part than on the surface part if the mating edge along the underlying part were recessed. The reviewing inspector or mechanic first noticing a surface defect should be suspicious and question whether or not any underlying parts might have similar or related defects. One useful technique is the *dashed line inquiry*. On most engineering drawings the outlines of underlying parts where two or more stacked parts are shown together are represented (when the underlying part edge is hidden from view or covered by the surface part) by a dashed line. On an actual packup of parts, it is helpful when looking at the assembly to visualize where the dashed lines should be. If the edge of the underlying part disappears from view at point A and reappears at point B, there must be some actual track of the edge between A and B. If this hidden edge is near a hole that is close to the edge of the surface part, the edge distance on the underlying part may be even more critical. Question and verify. On occasion it may be necessary to actually Xray the packup or use some other technique to either confirm or deny a suspected critical condition. Better to find a defect now than after completion and delivery of the product. (See Figure 6.2.)

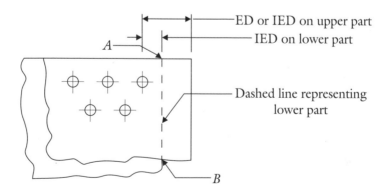

Figure 6.2.   The dashed line inquiry.

Other Parts Similarly Defective

Another part of the defect description is the determination and inclusion on the document of the number of parts or assemblies similarly (or identically) nonconforming, either by count and location if not serialized or by serial number if serialized. This is called the effectivity of the nonconformance and may require much further investigative effort to determine, especially if the defect was not specifically identified with a single unit. A generic type of defect (that is, one attributable to a design error, a worn or incorrectly used tool, or an improperly trained mechanic) may have occurred on many previous units, including some already delivered. In this case, a service recall or inspection may be required, depending on the severity of the defect and the anticipated consequences if left undiscovered and unrepaired. Happy indeed is the quality engineer who is able to determine with assurance that a particular nonconformance occurred only on the unit initially reported defective. Lack of such a determination may lead to later discoveries of similar nonconformances, either on parts already delivered, or parts yet to be completed. To put these thoughts into context it must be understood that seldom are all parts made inspected 100 percent of the time and to 100 percent of the requirements to which they were made. Statistical sampling inspection is common and generally reasonable. Exceptions to the norm can hurt.

Two sources for information on sampling inspection are MIL-STD-105E, Sampling Procedures and Tables for Inspection by Attributes and MIL-STD-414, Sampling Procedures and Tables for Inspection by Variables for Percent Defective. Either of these may be made a part of the contract. Instructions for their use could fill a separate volume.

Determination of Cause of Nonconformance

Related to the description of the defect, but not strictly speaking a part of it, is the determination of and the annotation on the nonconforming document of the defects true cause. As for the determination of the effectivity that generally follows the determination of the cause, the answers may be obvious but, more often, may be hard to come by. This is especially so since no one likes to admit error, certainly if punitive measures such as unpaid days off may result. Checks of documentation may help, but often the most fruitful avenues of approach involve personal low-key interviews with operational people and the implication, if not promise, of anonymity.

Often the determination of the true cause of a nonconformance may take substantial investigative time and many material review systems permit the completion of repair efforts while the determination of cause and the requirements for corrective action are still under way. Knowledge of the true cause of a nonconformance may be helpful in designing a repair, especially one that may be applicable to assemblies at various other stages of completion, but it is seldom necessary for a repair against the initial unit having the nonconformance.

7   The Review Cycle

Necessity for Review, Verification, Look, and Complete Description

The review of all the facts listed on the nonconformance document, or otherwise associated with it, is the necessary foundation for the building of a correct and effective disposition. The MRB engineer in so doing must not only verify the accuracy of the write-up, but become the technical expert relative to all aspects of the condition. All of the findings must be evaluated. All possible types of dispositions must be determined and considered and then action taken to permit the ultimate decision to be accomplished.

To do this, the MRB engineer must assume the role of the questioner or investigator, pursuing the answers to questions that may arise during the course of the investigation, adopting the role of (at least temporarily) a doubting Thomas and then verifying or refuting any doubts or suspicions. Among the most important aspects of the engineer's investigation is any possible physical examination of the actual nonconformance itself. It is almost axiomatic that the easier it is for the MRB engineer to examine the damaged part, the less complete will be the description of the nonconformance and its precise location. If the initiator of the nonconformance document is well aware that the reviewing MRB engineer can get up from the desk and walk 150 feet to actually look at the defect, the write-up may turn out to be somewhat sparse or abbreviated. The author has seen write-ups giving the location of a defect to be "as marked on part." If the MRB engineer were unable to examine the part, as is the case when the part may be several hundred miles away, the description would have to be clear and readily understandable unless the initiator believed that the defect would be fixed on the assembly, with no cost to the originator.

Defect descriptions written by a manufacturer of detail parts, who will only get paid for those parts when they are deemed acceptable or at least bought off by an MRB, will be substantially more clear than those initiated at a location where the reviewing MRB engineer is in residence. It doesn't take too many shipment delays (and the resulting missed delivery payments) due to the MRB engineer's inability to comprehend a cloudy defect description to bring about a substantial improvement in the clarity of the nonconformance write-up. The fact that this clarity of description often doesn't exist where the MRB engineer and the quality engineer are on the scene, suggests that local management may not know of the delays in disposition so caused, or may consider them less costly than hiring people who can come up with clear, correct, and concise nonconformance write-ups, someone like a junior engineer or draftsman.

Study Latest Engineering Drawings and Stress Reports, Determine Significance of Both Part and Nonconformance

At the same time as (or even before) the visual check, the reviewing MRB engineer should obtain a copy of all pertinent engineering drawings and unincorporated engineering changes and then study them in depth to verify that the engineering criteria against which the nonconformance was written are correct and up to date. The inspector may be excused for not having the most up to date blueprint, but the MRB engineer cannot. That is how one becomes an expert.

In addition to the blueprint review, the MRB engineer must make an initial judgment call as to whether or not the part containing the defect is of structural significance (that is, part of the primary load-carrying structure), or is what is commonly referred to as secondary structure, thus not necessary for the safe functioning of the final product within which the part is to be (or already is) installed. Having decided whether the structure is primary or secondary, the MRB engineer must judge whether the defect is of substantial or minimal significance. Experience and a questioning attitude are most important.

In the case of primary (and some secondary) structure the MRB engineer must find out if a stress analysis for the part in question exists and, if so, review it in detail to obtain a feel for the effect of the nonconformance on the strength and life of the final product. Suggested stress analysis techniques are discussed in Chapter 10. If the MRB engineer does not have a suitable background in structural analysis, an appropriate member of the design stress department must be consulted as an advisor to the MRB. Not all stress analyses undertaken are printed and published. Close relationships with the stress department are helpful here.

Determine True Purpose of Part, Verify Cause of Defect by Checking Other Parts

A result of these studies should be an understanding of the precise function or purpose of the defective part. Every part has its purpose and the MRB engineer must determine or discover this purpose if there is to be a correct disposition, or perhaps the more correct among several acceptable possibilities. The key question the MRB engineer must ask herself is "What does this part do?" Does it stiffen up the crew's thermos bottle holder against rattling? (If so, it is a secondary structure.) Is it an attachment fitting between the wing and the fuselage? (Clearly this is primary structure if the wing is not to fall off.)

Among the purposes of this review is the determination of the true cause of the nonconformance. Such determination may be assisted by undertaking some specific steps not always suggested by a read-through of the nonconformance description. For example (especially if the defect is readily visible), look at both the left- and right-hand sides of the assembly, even though only the left side was noted as discrepant. Of course this only applies if the right-hand side is supposed to be a mirror image of the left-hand side, or at least very similar in design. You may be surprised at what you find. Additionally, it may be helpful to look at the same location as the defect on earlier or later parts. Again, you may be surprised at what is discovered.

It also may be revealing to find out the configuration of detail parts that make up the assembly within which the original nonconformance has been uncovered. The assembly error may be caused, not by mislocated holes within the assembly, for example, but by individual parts within the assembly which are themselves too short.

A check of the accuracy of these detail parts, occasionally by the MRB engineer, but more often (and more officially) by the inspection department, may uncover the real reason for the assembly error. Such a check, called a *stock check* or *stock sweep*, can go a long way toward eliminating the recurrence of assembly nonconformances when the bad detail parts are detected early enough in the manufacturing cycle.

Is Part Interchangeable or Replaceable?

Another question to be asked during this review period is whether the parts containing the nonconformance have a requirement for interchangeability or replaceability (see Chapter 3). This has a major effect on the nature of the disposition since no repair or use without repair can be considered if the affected part or assembly cannot be interchanged from one unit to another.

The Personal Interview

Lastly, but among the most important of the review techniques, is the use of personal interviews of people having anything to do with the manufacture or installation of the defective parts. This is discussed in some detail in the last section of Chapter 6. If the MRB engineer is low key and adopts the attitude of being helpful rather than argumentative, and lets the individual talk freely at will, rather than (at least initially) answering direct questions, there is no limit to what may be uncovered. Obviously, there must be no hint or assumption of retribution or punishment. A tool engineer may reveal that the tool had not been checked after having been dropped several months before. The assembly mechanic may say that his partner's toe was broken because he wasn't wearing safety shoes when the tool was dropped on his foot. The inspector may mention that budget cuts had forced them to perform less inspection on detail parts. The mechanic may say that the part wouldn't fit properly unless it was twisted a little during installation. This part of the MRB engineer's job may be the most fascinating. Sometimes you get to feel like Sherlock Holmes. Other interviews with fellow MRB engineers who may have worked on the same assembly at other times, with the original designer, the stress analyst, the quality engineer on the job, and the supervising inspector may all yield pieces of the puzzle and may get you away from the desk and into the real world.

Checklists sometimes are provided for the MRB engineer as a reminder of questions to be asked and tasks to be accomplished to provide greater assurance for the accomplishment of a correct disposition. Each company's response to the request for such guidance by the neophyte MRB engineer varies, but the author (in response to such a request) came up with a set of MRB engineering guidelines for use by the MRB engineers within his company. (See articles Chapter 1, pp. 6 and 8.)

8 Disposition Possibilities

Variations on Remove and Replace

If every MRB engineering disposition was to remove and replace or to remove and scrap the nonconforming part, there would be no need for the MRB engineer. Replacement by a rubber stamp would be a cheaper alternative if the cost of the lost parts and the effort required to disassemble them were not also taken into account. If the cost of the individual defective part was small and all defects were discovered before assembly, the obvious course of action would be to discard and replace, especially if a correctly made replacement part were available without appreciable delay.

Unfortunately, the realities of the manufacturing world, cost, and time require that a disposition for nonconforming parts other than scrap and replace be a consideration. There are times, of course, when remove and replace or scrap and replace is the only possible action, when any other action would result in the use of an unsatisfactory part, a part that would fail in service or at least not properly fulfill its function as part of an operating assembly.

The terms *remove and replace* and *scrap and replace* have been used somewhat interchangeably, but there is a difference. Scrap and replace means that the defective part can serve no other purpose and must be destroyed. The replace portion implies that the replacement part must not have the same defect as the part to be scrapped. To be more specific, the disposition should probably read "remove and destroy part and replace with correctly made part," but the average MRB engineer is seldom this verbose.

The term remove and replace is even less specific and implies that perhaps the bad part could be used for some other purpose. If this is so, the MRB engineer should state it.

In the early stages of a new production program there may be a need for a test piece, a demonstration (nonoperating) piece or sales aid, a tooling mockup (sample) or a red master, a nonfunctional part painted a distinct color such as red and used as a temporary filler or tooling component. In this case, the disposition should state "Remove and replace with correct part. Defective part not acceptable for production, but may be used for mockup or tooling aid," (as the case may be). At several companies such parts are required to be painted purple, indicating no good for delivery.

In cases where the defective part was manufactured by a vendor, the disposition might be to remove and return it to the vendor and replace with an acceptable part. The vendor might choose to use the part as an anchor for his new boat or as a training aid. Return to vendor makes even more sense if the vendor paid for the raw material from which the part was made. So a disposition to not use a particular defective part can be worded differently, the more specific it is, the better. As in the case of all MRB engineering dispositions, knowledge of the history of the part is most helpful in arriving at the best disposition. It is better to leave little to the imagination of others. Implications can be interpreted varying ways.

Return to Stock

Some parts are considered defective because they do not mate properly with other parts, but may or may not themselves be defective. This can be the case when holes are drilled in each of two mating parts, instead of in only one of the two parts. Only after joining should the holes have been transfer-drilled through the existing holes in the one part into the previously blank (undrilled) other part. In this manner the holes would automatically line up with both parts at their correctly required locations. The holes in both parts may have been individually correct, but when the parts were located against other parts or clamped into a holding fixture (tool) they may have been mismatched enough to be unacceptable.

In this case, the disposition might be "Return to stock for later use with blank mating parts and attach to part a special tag stating, Consult document XYZ before use." Thus, the cost of the part is not lost. The MRB engineer must be assured, however, that the mating parts will become available and that other people will be given the task of ensuring that this takes place. Hardly the perfect disposition, but if properly followed through, an acceptable one. There are many variations on this theme; in other words, don't use this part now, but maybe it can be used later or for some other purpose. It is up to the dispositioning MRB engineer to determine the circumstances and make as clear a disposition as possible.

In some instances, later personal intervention may be required to ensure that the disposition is properly understood and carried out. The extra cost required for these later manipulations must be considered, however, along with the risk that the oddball part will be pushed to the back of the shelf in the stockroom and kept for the last vehicle in the contract when special mating parts may no longer be obtainable.

The Meaning of Use-As-Is

The best disposition of all from the standpoint of manufacturing and accounting may be use-as-is. This disposition may require many hours of investigation and numerical calculations (analysis) before the decision can be made, and the time required may not be worth it for a low-dollar-value part, especially if correctly made replacement parts are already in the manufacturing pipeline. However, it looks good, makes any shop error seem insignificant, and makes the MRB engineer look like a hero.

An unfortunate aspect is that manufacturing equates a use-as-is disposition with no more than ten minutes of prior research by an MRB engineer. Many an individual has asked why it took two weeks to come up with a simple use-as-is, not realizing that fifty pages of calculations were required along with extensive load determinations and tolerance studies. Considering the possible alternative dispositions, however, the cost was eventually determined to be well justified.

The words *use-as-is* mean just that, not do what you have to to make it fit or put it back on the shelf because another part fits better, although the use-as-is disposition doesn't necessarily prevent this from happening. It means that no unusual steps should have to be taken to use the part. If they do, the disposition is wrong. It also means that the defect should have no adverse effect (that is an effect beyond the allowable range of consequences) on either the fit of the part to any mating parts or on the purpose or function of the part. Use of the part may require the addition of a .090-inch-thick filler or shim when the drawing allows a shim up to (but not beyond) a maximum thickness of .090 inch. If the part requires an .091 inch thick shim, however, the disposition cannot be use-as-is and instead should have stated "use part but shim up to .091 inch using drawing-specified shim material." In the real world, most MRB engineers would probably still say "use-as-is" and most mechanics would not report the required use of a shim only .001 inch beyond the legal upper limit. People stretch limits to avoid unnecessary work, legal or not. If the required shim had been .125 inch instead of .090 inch, one could expect proper attention. In most cases a .091 inch shim would be acceptable, if not quite legal. If it were critical, engineering should have made a special notation of the applicable drawings using words such as absolute or mandatory.

The use-as-is disposition also does not guarantee that the originally designed strength or life of the part is the same as for a perfect or within tolerance part. It does ensure that the part maintains its strength within the allowable range of strengths and that its life will be no less than the minimum allowable life provided for by the contract under which the part and its deliverable end product were ordered.

If the defect was that the part was painted the wrong shade of gray (a cosmetic defect), then the strength and life of the part would not be affected at all and the use-as-is disposition would mean that the difference in coloration is acceptable without repainting. However, if the same part were supposed to be of .115 inch minimum thickness and was manufactured to a thickness of .105 inch, then both the strength and life of the part might be reduced because less material (for example, aluminum, steel, titanium, fiberglass) is working. This is especially true if the defect was located in the least-strong area of the part. Most parts are not uniformly strong or do not have the same stress level throughout. A reduction of strength in the strongest area rather than the weakest area may not lower the overall strength below the existing acceptable minimum strength of the part elsewhere. This would be like a chain with different-sized links. A weakening of the strongest link may still leave the chain stronger in the weakened area than in the nearby weak-link area. If the .105 inch occurs in an area designed merely for ease of manufacturing, rather than as a primary load-carrying element, the strength and life of the overall part might not be affected.

Static Strength and Acceptable Margin of Safety

In pursuing the meaning of use-as-is even further, some comments on strength and life are in order. The two primary measures of the ability of a structural-load-carrying member (that is, a member designed to resist a particular load applied to it during the life of the end product of which it is intended to be a component) are its static strength and its effective life (also called its fatigue life). The static strength is a measure of its ability to resist without failure or excessive stretch or deflection the maximum load expected to be applied to it during the course of its life. This load generally is increased by some amount for calculation purposes, thus providing a safety factor, the amount of the increase being dictated by the contract. This enlarged load (called the ultimate load) then is designated the design load.

Calculations undertaken by the stress analyst to determine the strength of the part result in the determination of the applied stress that the part is designed to handle—the design stress. Assuming a simplified example, this stress is the design load the part is expected to resist (measured in pounds, generally) divided by the cross-sectional area of the part (thickness in inches times the width in inches or the area in square inches), which is the area at right angles to the direction of the load. Thus, the design stress of a member having a cross sectional measurement of 1.2 inches (1.2") x 2.0 inches and subjected to a design load of 200,000 pounds (which includes the previously mentioned safety factor, not to be confused with margin of safety) equals:

$$\frac{200,000 \text{ pounds}}{1.2 \text{ inches} \times 2.0 \text{ inches}} = \frac{200,000 \text{ pounds}}{2.4 \text{ square inches}} = 83,333 \text{ psi}$$

also written as 83,333 pounds per square inch. This is the design stress for the design load (also called the static stress) resulting from the application of the static load.

If the member under consideration is made from a steel capable of resisting a stress as high as 125,000 psi, then the member is stronger than necessary. The amount of overstrength is quantified by use of the term margin of safety (MS). The MS is a measure by which the allowable stress exceeds the applied stress. In our example the MS equals:

$$\frac{\text{allowable stress}}{\text{design stress}} - 1 \quad \text{or} \quad \frac{125,000}{83,333} - 1 = 1.50 - 1 = +.50$$

a positive and thus desirable margin of safety. The allowable strength exceeds the design strength by 50 percent since the MS = +.50. A negative MS would indicate that the part was calculated to be understrength and would have to be enlarged. For the usual contractural purposes, a margin of safety of zero or above is acceptable (when considered by itself) and any defect which does not adversely affect the operational function of the part but reduces the design MS to no less than zero (excepting unusual circumstances) is acceptable. Continuing with the simplified example and assuming that the thickness of the part in the area supposed to be 1.2 inches actually measures .81 inch, the stress on the loaded cross section now would be:

$$\frac{200,000}{.81\ (2.0)} \quad = \quad \frac{200,000}{1.62} \quad = \quad 123,457 \text{ psi}$$

and the resulting MS would become

$$\frac{125,000}{123,457} \quad - 1 \quad = \quad + .012$$

a positive value. Under these circumstances, a use-as-is disposition would be acceptable as long as the calculated life of the part (as reduced by the thinner cross section) is not below the acceptable limit.

Service Life and Scatter Factor Must Be Acceptable

In addition to an acceptable static strength a defective part also must exhibit an acceptable life, requiring a different set of considerations and calculations. This subject will be covered in much greater detail in Chapter 11. If the defective part is not subject to substantial frequently repeated loads, it would not generally be considered fatigue critical or fatigue sensitive and the calculations to determine its useful life (fatigue life or life before the occurrence of a cracklike failure) would not have to be undertaken. Many parts used in aircraft type structure are not designed for fatigue and such parts can be acceptable for use-as-is as long as the MS is not negative.

For those parts, however, that are determined to be fatigue critical or fatigue sensitive the additional calculations must be undertaken. These calculations require not only knowledge of the applied loads, but the estimated number of times and the order in which they are applied to the particular part under review. The calculations will result in the number of hours of anticipated failure-free use, or the number of cycles or applications of the applied load that can be sustained without failure. See Chapter

3 for a discussion of fatigue critical versus fracture critical, a similar phenomenon that also requires additional calculations for remaining life.

The required life generally is stipulated by the contract as either the minimum number of operational hours (such as 10,000), or the allowable number of cycles (such as 2,500). Regardless of the calculated number of failure-free hours the correctly made part had been determined to be good for, a reduction in the number of hours as calculated for a defective part would still be considered acceptable as long as the resulting reduced number of failure-free hours or cycles was no less than the contract requirement, as sometimes modified for the use of nonconforming parts.

For fatigue-controlled parts, a term similar to that used for static strength (the MS) has been developed and is called the Scatter Factor (SF). (See Chapter 3.) When the calculated scatter factor for the nonconforming part is no less than that either stipulated within the contract or agreed to as the result of negotiations between customer and manufacturer, the nonconforming part may be accepted for use, subject to any special signoff requirements. Thus, for a program where the required operational life is 10,000 hours and any nonconforming parts must have a calculated life of at least 40,000 hours (having a SF of 4.0 or $\frac{40,000}{10,000}$), a defective part having a calculated reduced life of 42,000 hours would be acceptable for use, as long as the static margin of safety (which also must be calculated) is not less than zero. A positive MS is no guarantee of an acceptable SF and vice versa. Both must be determined when the part was originally or has become (perhaps as a result of an extreme defect) fatigue sensitive. The nonconforming part must be structurally acceptable, both from a static strength standpoint and from a fatigue life standpoint, when applicable.

Use-As-Is Rationale Statement

An additional requirement imposed on some MRB use-as-is dispositions is to state, as a part of the disposition, the rationale leading to the disposition. This can be a substantial aid to anyone questioning the details of the nonconformance or the reason for the use-as-is disposition at some future date. Unfortunately, some MRB engineers will use a standard statement such as, "Form, fit, and function not affected," which tells very little. Much more desirable would be a brief statement of what efforts were expended by the MRB engineer to determine the "use-as-is" disposition. Sample statements include, "tolerance study shows extra gaps still within shimmable limits," "stress analysis shows positive margin of safety," "defect cosmetic in nature only." The more specific the rationale the better. There is no benefit in hiding any heroic investigative efforts on the part of the MRB engineer.

Document Change Consideration

An addition to any requirement for a statement of rationale when a use-as-is disposition is employed is a requirement from MIL-STD-1520C applicable to those contracts invoking this specification. This mandates that the MRB engineer specifying a use-as-is disposition make a determination of the possible need for a document change (appropriateness is the word actually used) and so state this recommendation on the MRR. The MRB engineer may either state "no document change required," or "change recommended to drawing ABC or specifications XYZ, etc." When invoked, the reasoning behind this requirement is that a continual use of a use-as-is disposition may show that the condition is entirely acceptable for all time and that the drawing or specification particulars should be amended or relaxed proportionately.

Repair As Directed

The final disposition possibility covers the broadest category of all, *repair as directed*, and then spells out the necessary details for the repair to be accomplished. This possibility requires the talents of not only a stress analyst, but even more important, a structural designer; or in the case of a manufacturing facility specializing in electronic or hydraulic equipment, an electrical or mechanical engineer. Repair possibilities will be discussed in greater detail later, but some generalizations are pertinent at this point.

The repair instructions should be as complete as necessary for the manufacturing, tooling, and inspection personnel involved to correctly accomplish the work. This would involve reference to the same procedures, materials, and specifications necessary to accomplish the original manufacture. The disposition instructions should include all the necessary information, such as detailed layouts and/or fully dimensioned sketches to permit the fabrication of any special parts. In many cases the use of a copy of part of an existing blueprint (marked up by the MRB engineer) is very useful. The engineer should not leave out any information on the assumption that the mechanic will ask him personally for any missing data.

In some instances, it is acceptable to define a part as a slight modification from an existing part, spelling out the specific differences. In other cases, it is acceptable to define the shape of a part by reference to a required edge distance from existing holes the special part is intended to pickup. In all cases, it is recommended that the repair instructions be written in a step-by-step chronological order, describing first any tear down or removal operations required and any inspection measurements to be taken, then the details necessary to fabricate any separate or special parts, and lastly the assembly or reassembly instructions required to join all parts together to the reinforced configuration. In cases where the original part or assembly of parts was required to be tested in some manner, the repaired part or assembly also must be tested in the same manner.

9   Repair Possibilities, Types of Loadings

Repair Disposition Based on Strength and Economic Considerations

An MRB engineering disposition to repair rather than remove and replace an unacceptable nonconforming part usually is the course of action to take when the cost of removal is prohibitive as measured in either dollars or time. Given enough of each, anything can be replaced, but since many defects are discovered only after the defective part is extensively and permanently joined to many other parts, the cost and time for removal may not even be a remote consideration. Few structural parts are designed for easy removal and replacement; the weight penalty for such (particularly in the aviation/space field) is prohibitive. The fact that MRB engineers have been able to design suitable repairs to be accomplished in place, (on the assembly line, so to speak) has indicated that removal and replacement is not often required. If repairs were either impossible or not allowed by contract, the entire picture would be changed with much greater costs generated up front to ensure that assembly line defects did not take place. Tooling and inspection budgets would have to increase by orders of magnitude, an expenditure difficult, if not impossible, to amortize over the life of a low-volume production contract.

An MRB engineering decision to repair rather than use a particular nonconformance as is must be made when the nonconformance without repair either prohibits the defective part from acceptably performing its intended function, or makes it too weak (structurally inadequate) to either sustain the design loads without breaking or to continue in service for its full required operating life.

Capability of Accomplishment—Question Others

A disposition to repair generally is based on economics and the ability to design a structurally, operationally, and cost-effective repair is the most important talent the MRB engineer can possess. Cost effectiveness must be a measure not only of the dollars saved by not having to manufacture and install a correctly made replacement part, but the capability of manufacturing and correctly installing any necessary separate repair parts. A repair part that turns out to be virtually impossible to install may be less cost effective than removing and replacing the damaged part. Experience and first person communication with the actual individuals expected to accomplish the repair are among the keys to the design of a cost-effective repair. One of the most disheartening comments an MRB engineer can hear from the manufacturing side of the house is, "I wish I had known how complicated this repair turned out to be; it would have been cheaper and faster to have replaced the part."

To avoid this possibility, some initial research is recommended, especially when the MRB engineers' prior experience with a particular type of repair and location on an assembly is minimal. Among the most obvious recommendations is to query others for suggestions, in order to locate any previous repair dispositions against the same or similar parts. Another goal is to solicit ideas. Unless a part or assembly is among the first few manufactured, there may exist a previous repair that is either directly applicable or amenable to suitable modification. The mechanic and inspector most closely associated with either the discovery of the nonconformance or the manufacture of the part should be asked if they have seen the problem before and, if so, what happened. If not, they both might be asked if they have any ideas on how to fix it. This serves to make them a part of the repair action, as well as to introduce the desire to gain from their experience.

An MRB engineer too bashful (or considering himself too professional) to ask these questions may be cutting off a valuable source of information, helpful not only to the present problem, but for possible future use. Some of the author's best ideas have come from other people's suggestions.

Another source of information on prior repair dispositions is another MRB engineer, a quality engineer, or a design engineer who may have been associated with a similar repair on a prior unit. This particularly makes sense when the MRB engineer is newly assigned to an ongoing job. Pick the brains of those who preceded you. They may be more than happy to offer both suggestions and assistance, since well-thought-out and successful repairs can be a measure of pride to an accomplished engineer.

Research Prior Repairs

Another source for the discovery of a prior applicable repair may be a data file maintained within the MRB engineering office. Some offices maintain a list of part numbers repaired and the nonconformance document numbers relating to each repair. The use of computer logging systems permits relatively easy retrieval of these numbers, although the tendency to list documents against the assembly within which the discrepant part is located, rather than against the actual discrepant part itself, tends to dilute the benefits of this approach. A search through 50 MRRs against a major assembly in hopes of finding one particular repair can be discouraging, as opposed to reviewing two MRRs against one particular detail part, but it does have the benefit of exposing the neophyte MRB engineer to a lot of repair design ideas.

The MRB engineer must be very cautious, however, when considering the use of an already accomplished repair on a new defect, even if identical to the original defect. The use of such repairs without careful consideration as to their precise applicability to the current condition, is fraught with danger and must be considered, at least initially, as a possible starting point only. Conditions change and what may be suitable in June might be unacceptable in September. An in-depth review must still be conducted. Key personnel needed to accomplish the repair may have relocated or retired, loads expected to be applied to the structure may have been found to have increased, thus requiring a stronger repair. In some cases the configuration of the structure itself may have changed. A small sheet metal clip may have been added to carry a length of electrical wire, thus getting in the way of last year's easily applied reinforcement. Beware of placing blind faith in a predecessor's work.

The old repair, however, may be a good starting point and, suitably redesigned or modified, could save the MRB engineer a lot of exploratory time. She should realize though that once her signature is in place, she bears full responsibility for the design of the repair. The argument that, since Joanne did it, it must be okay, doesn't hold water.

Modify Standard Repairs

Another source for inspiration when seeking repair possibilities is the standard repair manual. These repairs tend to be generic in nature and can often be applied to numerous defects, wherever located, as long as any required added parts can be physically accommodated without sacrificing other unique requirements the structure or equipment may have to sustain. As in studying old repair designs, a look through the standard repair manual can show how it is done. In some cases the standard repairs are limited in application, but when specified as written or, with any necessary modifications, as a disposition for the full MRB to consider, they can be used for more extensive applications when the actual structure on which they are to be used can accommodate them. For example, the limitation on a standard repair that no more than two enlarged holes in a row can be repaired with oversized rivets might be waived when four enlarged holes in a row exist in a location used only as a seal against dust.

Must Know Significance, Precise Purpose, Strength, and Life of Defective Part

The matter of determining the applicability of a particular repair or the need to modify it often requires the knowledge of the structural significance of the part having the nonconformance, along with a determination of its structural margin of safety (see Chapter 8) and its calculated fatigue life. The selection of a physically possible repair often is easier than determining the resultant strength and life of the damaged part with the repair reinforcement riveted in place. This requires some knowledge of stress and fatigue analysis, or at least the ability to recognize the need for such an understanding and the need to consult others who are more knowledgeable.

An ability to differentiate between various repair possibilities and provide for the most effective (read inexpensive) repair requires some knowledge of what the defective part is doing (in other words, its purpose). This applies as well to equipment, wiring, fluid-carrying members such as aluminum tubing and whatever other device may be defective as to structural elements or members. The repair must restore the function of the part lost or reduced by the occurrence of the nonconformance. As previously mentioned, it may not need to restore it to 100 percent of its original value if the part already was overdesigned, but the part as repaired must meet its required purpose. Of course, if a fuel line has a leak, all of the leak must be stopped, not just most of it. In this case, restoring the function of the part to 90 percent of its original purpose is inadequate.

Structural Elements—Compression, Tension, and Bending Loads

The following discussion applies to structural elements, those designed to carry the forces or loads expected to be applied to the product throughout its active life. For example, typical structural elements for a building would include its beams, columns, joists, rafters, lintels, girders, footings, and so forth. These members are expected to sustain a weight or force (often expressed in pounds) that may be applied vertically, horizontally, or both, at one or many specific locations or spread out along a measured length. The loads expected to be sustained by these members can be determined, and the ability of these members to resist these loads without falling can be calculated. This is the province of the structural engineer with knowledge of structures and stress analysis, gained through education and experience.

All structures have load-carrying elements, although the loads on many small structures (such as appliances) are insignificant compared with the strength of the members required to give them form and function. Larger structures, however (such as vehicles), share many things in common with buildings. They all have structural elements such as beams, columns, and shear webs.

For simplicity we can define a beam as a long, relatively thin, member that is designed to carry loads acting at right angles to its length and thus bent downward at its middle when held from falling at its two ends. This assumes that the loads are acting downward, as under the force of gravity, and that the beam is positioned somewhat horizontally. It also assumes that the member is supported at both of its ends. Beams are not always horizontal and not always restrained at both ends, as with a swimming pool diving board. The floor you walk on at home is supported by beams called joists if there is a room, basement, or crawl space below. Most beams are installed level or horizontally at the time of initial fabrication.

A column (or post) can be defined as a short or long, relatively thin, member designed to carry a load pushing against it in the same direction as its length and thus tending to compress it or make it shorter. The load it experiences is called a *compression load* and the effect of this load acting on the column is called the *compression stress*, actually the number signifying the apportioned amount of the load applied to each square inch of the part of the column (its cross section) resisting the load. The numerical result is expressed in terms of pounds (of load) divided by the area of the column (square inches) or pounds per square inch, commonly referred to as psi.

Thus, a column measuring 4 inches x 4 inches (a square column) has a cross-sectional area of 4 x 4 = 16 square inches. When supporting without failure a load of 19,200 pounds, it will sustain a compressive stress of 19,200 divided by 16 = $\frac{19,200}{16}$ or 1,200 psi, a miniscule stress for most metals, but a substantial stress for most woods.

This discussion is purposely avoiding what would happen under this loading if the column were extremely long, as this introduces another problem, the tendency for long, thin columns to flex sideways or buckle. In addition, columns made of thin-walled elements of open sections such as angles, zees, or channels may be subject to local buckling and crippling failure, depending on the specific geometry of the column's cross section. The MRB engineer is advised to seek stress department assistance if necessary and if this is a consideration.

The column is a prime example of a compression member, a structural element experiencing primarily (or solely) a compression load. If the same member were to sustain the same load, but in the opposite direction, it would tend to be stretched or lengthened. The load would be called a *tension load* (generally considered to be a positive or plus (+) load) and the 1,200 psi stress would be called a *tension stress*. The member would not be called a column, but a tension or tensile member, or often an axial member; the term axial referring to the predominant load direction along the member's axis.

A column also could be called an axial member, but seldom is. A beam is a bending member (when subject to bending loads) and the stresses sustained by the beam under the action of the bending load are called *bending stresses*. They are calculated in a substantially different way than tensile or compressive (that is, axial-type) stresses. A failure under any of these loads is called a structural failure, but the member to which they are applied, if properly designed and manufactured, will resist these loads without failure for as long as it is in use and experiencing the loads, or at least as long as it was designed to carry them (which might not be the same). This discussion is somewhat simplified, of course, as many structural members are meant (designed) to carry more than one type of loading, either separately or together, and the loads themselves may change in a random or cyclical fashion. This complicates the life of the structural analyst, but makes for more efficient use of the structure.

Shear Web Defined—Shear Stress and Shear Flow

Another type of structural member is the shear web, harder to define, but no less necessary in modern structures. In olden times, most wooden buildings were of the post- and beam-type construction with horizontal beams carrying the loads from above to the outer walls where the beams were supported by vertical posts (columns) extending downward to the ground and some type of foundation. The walls between the posts were not considered as load carrying members. In more modern structures, the wall-like elements between the posts (when rigidly secured to not only the posts but to structural members along the other two sides of the rectangular shape) are called *shear webs*.

The term *web* implies that the member is thin relative to the bulk of the supporting members on all four sides. The term *shear* describes the process whereby a load introduced along one side of the rectangular (or trapezoidal) web transfers across the web to the other side in a sliding or shearing action within the web's plane. (See Figure 9.1.)

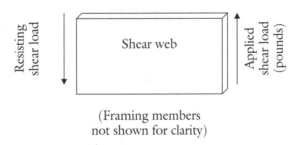

(Framing members
not shown for clarity)

Figure 9.1. Shear web.

The web must be suitably fastened to its framing members along all four sides for this to happen. If the thickness of the web is large enough to sustain the shear load that passes across it without flexing, wrinkling, or buckling, the web (sometimes called the shear panel) is said to be shear resistant and sustains a *shear stress* defined as the load applied along a side of the panel divided by the area of the panel along that same side. Shear stress is then equal to the load per square inch of web resisting that load and parallel to the direction of the load. A load of 1000 pounds applied along (not at right angles to) a six-inch length of web having a thickness of .030 inch will introduce into the shear web a shear stress of

$$\frac{1000}{6 \times .030} \ = \ 5556 \text{ psi}$$

Sometimes it is convenient to highlight the load along each inch of the six-inch length of the web at this point. This would equal 1000 pounds divided by six inches, or approximately 167 pounds per inch; a new term described as the *shear flow* applied to the shear web along this particular side or edge of the panel.

For shear webs that are of unequal length along all four sides, the shear flow will be different along each side or edge of the web. It should be noted that under certain circumstances a shear web, especially when supported by substantial edge members, can carry without failure a much larger load than that of a magnitude to keep it from buckling. This larger load will cause the web to wrinkle in a manner suggesting diagonal furrows in a plowed field, but the web and the surrounding frame members still will not fail. The web is then described as having gone into diagonal tension, a subject beyond the scope of this chapter.

Shear Loads on Axial and Bending Members

The term *shear* and *shear stress* have been described as applied to a thin rectangular web secured on four sides to relatively more substantial framing members. Shearing loads also can be applied to axial-type members carrying tension and/or compression loads. The difference is that the tension or compression loads would result in some (however small) lengthening or shortening of these members, whereas a shear load would be applied at right angles to the tension or compression loads and would have a tendency to slice the members in half as if a huge pair of shears were placed into position.

Many structural members carry not only axial (tension and/or compression) loads, but shear loads as well. The most common exception to this is found in a structural truss whose members are designed (sized and positioned) to carry only axial loads. Beams, however (bending members), carry primarily shear loads causing them to bend away from the loads.

Structural Analysis Capabilities or Assistance Required

A knowledge of the types of loading conditions that a structural member is designed for (its purpose) is necessary to permit an effective assessment of its importance to the assembly in which it is to be installed. For this purpose a background or some training in structural analysis is most helpful. Additionally important is the ability to determine or obtain the actual loads expected to be applied to the member (its design loads) and to be able to undertake a stress analysis of the part under consideration. This requires a background in stress analysis. The combination of structural designer and stress analyst is hard to find, but is a valuable commodity in the MRB engineering field. In the real world the MRB engineer from the stress department may not necessarily be a good designer and the MRB engineer from structural design may not make a good stress analyst. The possibilities depend on individual temperament, capabilities, and the job tasks to which the engineer is assigned, coupled with the abilities of her mentor on that particular job. Opportunities to perform both structural repair design and stress analysis must be made available to the neophyte MRB engineer if full potentials can be realized. If the MRB engineer cannot do any necessary stress analysis she must (once she recognizes the need for stress analysis on a particular nonconformance review) seek help from one who can, either within the local community of MRB engineers or from the company's stress department. Cross training is highly recommended when the needs of both the MRB engineering and the stress departments can accommodate the temporary switch in personnel.

Doublers

Once an analysis has shown that a defective part cannot be used as is because of a resulting understrength condition, the most obvious choice is to design and have made and installed a suitable reinforcement. Such a reinforcement generally is provided for by the design, manufacture, and installation of an additional part (an add-on), a part that can pick up some of the required load and share the burden with the weakened part. This requires that the added part be able to pick up some of the total load downstream of the defect on the nonconforming part (by means of suitable attachments), carry this portion of the total load across the weakened area (a bridging action), and then dump this load back into the original structure upstream of the defect. Since the reinforcing piece (often called a doubler) picks up some of the total load, the load remaining in the defective part at the point of weakness will be less and the resulting stresses (load divided by remaining area) will be no more than for an unreinforced but properly made part. In fact, if the doubler is large enough and the fasteners at the two ends of the doubler are suitably massive, the doubler will pick up an increasing percentage of the overall load, reducing the stresses at the weakened area of the original part even further.

The furthest extreme to this progression would be to make the doubler so heavy and so stiff that it picks up all of the original load. This could be guaranteed by leaving the defective part in place, but cutting it in half at a the weakened section. The reinforcement would then have to be stressed (undertake a stress analysis) to ensure that it could carry the full original load without failure and for the period of time required. The technique of adding a doubler is very common in the MRB engineering repair world and is similarly used by the structure design engineer when the loads applied to the structure are found to be larger than when the parts were originally designed. The doubler can be said to provide an additional load path and to reduce the stress level in the part being reinforced. The biggest problem with doublers is that they add weight, may overstress any existing fasteners used to install them in the load pickup and load dump areas, and must be installed in an area where there are no other parts already in the way. Doublers have been designed and made and then found to be incapable of installation because another part already was in place.

The term *doubler* tends to be somewhat generic in nature and probably was derived from the doubling up effect. The term *tripler* also has been used for additional add-ons. Care should be taken to distinguish between an added reinforcement which, in the form of a flat, skinlike member, is a true doubler, and a replacement part which is exactly that. A doubler is an add-on. Many other types of reinforcement may be employed that are not in the normally understood form of a doubler, but serve the same purpose. These may be in the form of angles, channels, or other shapes having legs, flanges, and webs rather than just the single sheetlike form of an original design doubler.

Sometimes these added reinforcements are merely identified by the type of cross section they have, such as an angle that has two legs of either equal or unequal width and thickness. Sometimes they are called reinforcements or straps. Each additional member can serve to bridge the weakened area on the original member and when properly designed can restore the lost load carrying ability of the nonconforming member. Generally, they are made from the same type of material as the original, but occasionally when space is limited they may be made from a stronger material. Steel doublers have been used against aluminum structure (the steel is approximately three times stiffer than the aluminum) and steel doublers heat-treated to a stress level of 270,000 psi have been used to reinforce steel members heat-treated to a stress level of 125,000 psi. Titanium doublers also are used where weight is critical. Since this is not a one-on-one type of reinforcement, the differing stiffnesses of the different materials must be taken into account when calculating load pickup capability. The stiffer material will soak up loads much more readily than the less stiff member having the same dimensions, and may fail if not suitably analyzed. Additional consideration must be given to protecting the dissimilar metals from the dangers of galvanic corrosion. One or both members should have a finish or coating applied that will eliminate, or at least greatly reduce, this danger.

Use of Existing or New Fastener Locations to Permit Doubler Load Pickup and Dump

One of the most important requirements for the successful use of a doubler or other reinforcement is the provision for the proper pickup and dump out of the portion of the original load that the doubler is expected to carry. In most cases, other than welded or adhesive bonded structure, this involves the use of fasteners such as rivets, bolts, or close tolerance pins of various types. The more common technique is to make use of existing fasteners. The less common, but often better choice, is to add new fasteners at locations where existing fasteners are not already in place. The choice often is dictated by available space and the configuration of the existing structure in the area of the defect. In some designs it is possible to make use of existing fastener locations and add or interspace additional fasteners.

In other designs the existing fastener locations may be used, but the original fasteners, once removed, are replaced with the next larger standard size or with similar size fasteners (at least as large as the original, but never smaller), but of a stronger material. Common larger fasteners have shank diameters 1/64 inch or 1/32 inch oversize, specifically designed for salvage or repair purposes, and 1/32 inch, 1/16 inch, or 1/8 inch larger standard diameters, depending on the range of sizes available from the fastener manufacturer or carried in stock in the assembly area.

Knowledge of fastener capabilities is a field in itself and the various fastener manufacturing companies are more than willing to assist with suitable literature, especially if substantial fastener sales may result. The significance of this statement becomes clear when it is realized that certain large diameter fasteners used in the aerospace industry may cost upward of $80 each with a minimum order of 100 fasteners expected.

The author was once associated with a program where the cost of purchased salvage (repair) type fasteners and larger size standard-type fasteners was much more than the cost of a single completed article within which the fasteners were to be installed. It took the completion and delivery of many more units before the cost of the repair fasteners was recovered. Fortunately, the original contract for two articles led to follow-on contracts for several hundred more over a period of about five years.

Doubler Fastener Load Requirements—Finite Element Analysis

Knowledge and use of stress and fatigue analysis and the more recently developed finite element analysis is necessary to correctly design the fastener attachment pattern for doubler load pickup and dump. Without getting too involved finite element analysis can be defined as that analytical technique that can determine the actual load that each separate fastener will experience within a group of fasteners designed to carry an overall load applied just before the start of the fastener pattern.

In addition, when the fasteners support more than two adjoining or mating parts, the amount of load that will be carried by each separate part along each individual fastener can be calculated; a powerful tool indeed in the hands of a dedicated MRB engineer.

The analysis that determines load apportionment among fasteners in a multifastener pattern is sometimes referred to as a *rubber bolt analysis* because the relative stiffness of the fasteners and the elements of the metal parts between each pair of fasteners are entered into the calculations, and the reciprocal of the word stiffness is the word flexibility, or rubberiness. For starters, we have found that the load on each of four fasteners carrying a total load of 400 pounds (400#) is not 100# or 25 percent each, a value that was assumed to be the correct load on each separate fastener for many years. In such a joint, where the four fasteners transfer the total 400# load from part A to part B, the load on the two end fasteners, the first and the fourth fastener, may be up to 30 percent greater (or 130#) than if the load were evenly distributed. The load on the two middle fasteners would be less than for an equal distribution. (See Figure 9.2.)

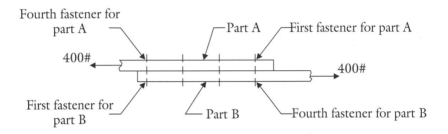

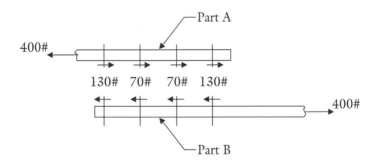

Figure 9.2.   Fastener load distribution.

Knowledge of the precise loads on each fastener is extremely helpful for proper doubler design. Such knowledge suggests that in using existing fasteners (actually they are removed and later replaced), they may be overloaded by requiring them to carry additional loads for which they were not originally designed. A newly located fastener does not suffer this indignity, but the placement of a hole in the previously blank area of an existing part may weaken the part itself, either because of the hole or by the introduction of a load where no load existed before.

Fastener Shear Load and Resulting Part Bearing Stress

Coupled with the ability of a fastener to carry a load between one part and another or, better stated, from one part across any gap (there should be none) to its mating part is the ability of the parts themselves to pick up or carry this load without failing. The fastener itself carries the load in shear across the joint. This load is called the *fastener shear load*, as if a giant pair of shears were attempting to shear the shank of the fastener in half. The shear load on the fastener is introduced into the fastener at the hole in the part within which the fastener is installed by the pressure of the part against the shank of the fastener. This pressure occurs against a rectangular area of metal pushing against the fastener and defined by the thickness of the part (t) along the axis of the fastener and the diameter of the hole (D). (See Figure 9.3.)

Shank of fastener in contact along
hole in part

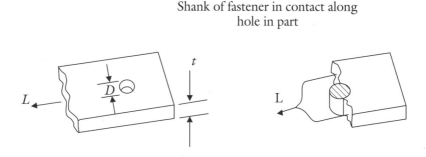

Figure 9.3. Fastener bearing pressure.

This pressure is called the bearing pressure against the part, and the *bearing stress* is the fastener load (L), (against that particular part at that particular hole only) divided by the wetted area resisting the load or L/Dt. Wetted means that area in contact between two parts, the area that would be wetted if one of the parts had wet paint on it at the time of initial contact.

Each type of material has its own allowable bearing stress, F_{bru}, where F stands for an allowable stress, the subscript $_{br}$ is an abbreviation for the word bearing, and the subscript $_u$ stands for ultimate. Values for F_{bru} are published in MIL-HDBK-5 as well as in many fabricators and manufacturers' data sheets. This allowable bearing stress must not be exceeded by the applied bearing stress. If so, the hole will become elongated in the direction of the applied fastener load (sometimes called the pin load), the joint will loosen up, and an eventual failure may occur. Thus, two criteria must be checked for positive results if a fastener joint is to be acceptable. The shear loads on the fastener shanks must not be larger than the allowable shear loads the fasteners can carry and the bearing stresses against the fastener holes in the parts being joined must not be larger than the allowable bearing stresses for the materials from which the parts are made.

Fastener Allowable Loads

Fastener shank allowables are published in MIL-HDBK-5 and other sources. They are based on statistically oriented test results.

For protruding head fasteners, the test straps must be thick enough to resist the bearing loads from the test beyond the point of the fastener shank shear failure, thus forcing the failure of the fastener shank itself. This is given as a single value, the maximum load that the fastener can resist regardless of how thick the test straps are, and is considered the allowable single shear load or fastener shear strength. Single shear means that the loading is applied through one sheet or strap only and resisted by one sheet or strap only, the specific point along the shank of the fastener between the two straps being the failure location or, more specifically, the single shear face of the fastener. If the load was applied through one strap, but resisted by two straps of equal thickness (one on each side of the single strap), the fastener could then dump its load to both of the resisting straps and the fastener would be considered to be in double shear; each of the two shear faces at a different location along the shank. The fastener could carry twice the load as when in single shear, as long as the single strap thickness was enough to carry this load. In some instances, a single fastener might pick up as many as a dozen individual parts, so that many individual shear faces would result.

Keeping track of the loads can get complicated under these circumstances and the MRB engineer must check the loads and directions of application of each to verify that the total of loads applied to each individual fastener equals zero. This is called a *static balance* and is akin to determining that the pull on both ends of a rope is the same if the rope is to remain stationary and not be heading north.

The determination of fastener allowables for flush head fasteners is more complicated. They require a tapered recess or countersink in the part within which the head is to be seated.

Failure of a protruding head fastener when the fastener is subjected to a shear load (more commonly experienced than a tension load) is based on the capacity of the shank to resist this shearing load, as long as the parts on both sides of the shear face are thick enough to not fail themselves. However, a flush head fastener requires that a portion of the material in the part that the head of the fastener is to be recessed in be cut away to accept the head. This action weakens the ability of this part to carry as much shear load as the fastener shank itself could carry before failure. The thinner the part in which the fastener is placed, the less load the joint can carry. This is not only due to the loss of material resulting from the countersink, but also to the prying action the countersunk part applies to the fastener's head.

The walls of the countersunk recess are not at right angles to the direction of the load, as the walls of a plain straight hole would be, so the tendency is to bend the head of the fastener. Fastener strength values for countersunk fasteners are given for various countersunk sheet thicknesses and types of material. Those above the thickness (and type of material) resulting in a test failure just equal to the fastener shank shear capacity itself will show an allowable load equal to the allowable shank shear. Those sheet thicknesses increasingly less than this equal strength thickness will show markedly less joint strength values. It also should be noted that the bearing strength of the countersunk sheet contributes to these values. An additional bearing resistance calculation is not necessary.

A knowledge of fastener/sheet combination load allowables is necessary for correct calculation of doubler effectiveness. Larger or stronger fasteners can pick up more load as long as the parts in which they are installed can carry the increased loads and are not themselves unduly weakened because of the larger holes required.

The Three-Rivet Plate

An analysis of the effect of adding a doubler or increasing the size of a fastener must include consideration of the effects on adjacent structure. A prime example of this is the three-rivet plate, a doubler designed to pick up the load from the endmost rivet (or other type of fastener) along a row of rivets. (See Figure 9.4.)

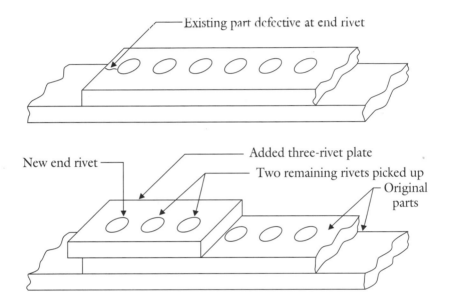

Figure 9.4.   The three-rivet plate.

The load from this end rivet, because of a defect in the part in which it is installed, is presumed to enter into the first rivet location of an added three-rivet plate, or strap of the same material, width, and thickness as the defective part. This load then, once in, must dump back out through the two remaining rivets. All three rivets at the original locations would have been removed and replacement (new) rivets installed, picking up not only the original two or more parts, but the added three-rivet plate as well. The problem is that the loads on the original part at rivet locations two and three are increased beyond the original because they now pick up the load from the plate. If the part seeing this local load increase is thick enough, it will still be able to resist the increase in bearing, but if not, it may experience a bearing failure. Any analysis of a three-rivet plate reinforcement must determine if the thickness of the original part is adequate to resist this increased load.

Consideration of this possibility suggests that it might have been just as effective to prevent the first rivet from picking up its share of the load by drilling an oversize clearance hole at rivet location one and letting rivets two and three (and all others in the same row) individually assume a larger share of the original total load. This is the type of thinking that is looked for from the more experienced MRB engineer.

Provide a Fastener Clearance Hole or Enlarge Adjacent Fasteners

The use of a designed clearance hole to prevent fastener load pickup at a particular location is an effective way to eliminate damage possibilities when the part within which the clearance hole is drilled is separately weakened adjacent to the hole. This action revises the load path. It forces the load to the remaining fasteners. Another way to accomplish this is to remove and replace the adjoining fasteners with larger fasteners. This changes the distribution of loads within the same pattern of fasteners originally designed to carry the load. The big ones carry more; the remaining smaller ones carry less. Repairs need not be limited to adding reinforcements.

Surface Damage—Blend Out, Calculate Remaining Strength and Life

It is frequently possible to remove material from a nonconforming part to effect a suitable repair. Many defects involve local surface damage to a part when the part is dropped, damage caused by an accidental drilling into the part from a nearby hole, or damage caused by a mishandled screwdriver resulting in a scratch or gouge. Most such types of damage should not be left in place since they may not be cosmetically acceptable but, more importantly, may have sharp edges which can lead to the development of a crack and eventually to a full-blown fatigue failure. A cardinal rule for MRB engineering repair dispositions is that unless very real extenuating circumstances exist, always specify the smooth blend out of all mechanical surface defects such as gouges and scratches.

One criterion for smooth blend out is that, after blend, no fingernail pickup should occur. The inspector's fingernail, when slowly drawn across the blended surface, should not be able to rest against or hit any ridge along the surface not removed by the blending.

The surface roughness after blend, which usually is accomplished by gentle abrasive methods, should be at least as smooth as required for the original part, preferably better. That is, for a part required to have a surface roughness of 125/ (see Chapter 4), it would be beneficial to specify a blend out surface roughness of 64/ or better.

Once the defect is considered as blended out, the effect on the stress level of the part due to the reduced area of the working metal remaining has to be evaluated. If the margin of safety (see Chapter 8) is higher than zero there is some excess material to play with and the material remaining may still be adequate to carry the design load. Calculations may be necessary to verify this. In addition to the effect on the static strength of the part, because of the reduction of load carrying metal, the effect on the fatigue life must be determined if this is a consideration of the original design. Fatigue life often is related to the configuration of a part at the precise location where a reduction of its thickness occurs. If the step on the surface, where a thickness is reduced from, say, .25 inch to .188 inch, is sharp or square edged, the fatigue life will be much less than if the transition requires a gentle radius or a smooth gradual slope.

Such a radius is a prime determinant in the calculation of the part's life. The larger the radius, the better is the rule of thumb. Thus, the gentleness of the blend can be very important. Cases exist where gentle blends to reduced thicknesses have a longer fatigue life than the original parts with a small radius between the original thicknesses. Elimination of the damage rather than repair or reinforcement may sometimes be the better way to go.

Damaged Fastener Hole Repair Choices

Much discussion has ensued on the subject of fasteners and their use in supporting added reinforcing members such as doublers. A more common use of fasteners as a repair possibility is in addressing the problems associated with damaged holes. As previously suggested, the enlargement of a damaged hole to clean out any sharp edges and reduce the possibilities of a fatigue failure due to the sharpness or the possible existence of an associated crack, should be just about mandatory.

The choice between (1) enlarging the hole beyond the minimum cleanout diameter to permit the correct tolerance installation of the next increment oversize fastener, (2) the installation of the blueprint diameter fastener in the oversize hole (minimum load transfer due to friction only), and (3) direction to not install any fastener at all, would be based on the requirements of the structure. A highly stressed fastener pattern would require a greater degree of structural analysis than would a joint where the fasteners were merely used for the positioning of nonstructural parts against one another. In some cases, a fastener would be required at this location to clamp two parts together to prevent a fuel leak or prevent water intrusion. In other cases, the fastener may be fitted with a sealing device such as an O ring, a doughnut-shaped rubber washer, or some similar sealing ring.

In this case, a fastener would be required along with an equivalently larger or oversize sealing device. Again, the bottom line is that the precise purpose or purposes of the original fastener must be known and this requirement not lost in the selection of the repair approach.

Deepened Countersinks with Knife Edges

When calling for larger diameter flush countersunk (not dimpled) fasteners, consideration must be given to the effects of the generally required deeper countersinks. Unless the oversize fastener has a small head (called a shear head), instead of the larger tension head of the original fastener, the countersink would have to be deepened. This can result in what is known as a knife edge, a phenomenon to be avoided.

When the depth of the countersink approaches or equals the thickness of the part in which the hole is located, a sharp circular edge of metal results at the intersection of the sloping surface of the countersink and the straight shank portion of the original or enlarged hole. This produces a chisellike edge, circular in shape, and is universally called a *knife edge*. (See Figure 9.5.)

No good results from this. The strength of the joint is reduced since most of the load now has to be resisted by the sloping surface and the fastener

Figure 9.5.   Knife-edged countersinks.

head has a greater tendency to be pried off. Additionally, the fatigue life of this type of joint is substantially less than for a protruding head fastener and less than for a similar flush head fastener with a shallower countersink. For these reasons it is highly recommended that the repair be designed so that the final countersink depth is no more than about 75 percent of the thickness of the individual part in which the countersink is located.

The dimple referred to in the previous paragraph, in which the member under the head of a flush fastener is physically pressed inward around the hole in a shape to receive the head of the fastener, is not subject to the same limitation. Dimples themselves however, particularly in high-strength materials, have other problems (such as a potential for cracking) and are not as widely used as in the past.

Repair Fastener Stocking

The use of a 1/64 inch or 1/32 inch oversize diameter fastener as a repair choice for a fastener hole that can be enlarged by these amounts to remove all damage and provide a clean hole within the necessary tolerance limits has been discussed at length. Unfortunately, the procurement and stocking of the necessary salvage fasteners is costly and must be done well in advance of need, considering the lead times (period between order from the fastener manufacturer and receipt of the fasteners) required. In addition, each separate length fastener of any particular diameter must be ordered and stocked since shorter fasteners cannot be used as a replacement for longer B/P fasteners, although longer fasteners can often be used if the increased lengths of the shanks do not hit adjoining structure. Added washers sometimes can be installed along the unthreaded portion of the shank of overly long, threaded, oversized fasteners, but the types of oversize fasteners (such as blind rivets) that are required to be of the precise length to match the total thickness of the packup cannot be used this way.

Use of Repair Bushings

Two alternatives are available to permit the use of standard diameter fasteners in oversized holes: specify the use of a repair bushing or specialized type of thin-wall bushing called an ACRES sleeve. A bushing is basically a collar used in conjunction with a fastener. It looks like a short length of pipe. The length of the bushing is designed to equal the thickness of the part or packup of parts into which it is to be installed. The inside diameter generally is specified to be the same as the diameter of the original hole for the B/P fastener and the outside (larger) diameter is designed to provide an appropriate fit against the oversized hole.

Sometimes an interference (controlled press) fit is specified, especially where fatigue life is a consideration.

The bushing fills the space between the shank of the reinstalled B/P fastener and the enlarged hole, transferring the fastener load into the original part, or housing, in which it is to be installed. Bushings normally are made of a material somewhat stronger than the housing material (the part surrounding the bushing) since the bearing strength of a bushing equals 1.304 F_{cy} (1.304 times the allowable compression yield stress of the material from which the bushing is made). That strength is less than the F_{bru} or allowable ultimate bearing stress of the same material. The yield stress of a particular material is that stress at which the material, under a test loading, starts to stretch or strain at a markedly more rapid rate than during the lower stress part of the test. The ultimate stress is that stress at which failure or rupture takes place. These values are given in MIL-HDBK-5.

Since the allowable bearing stress that a bushing can resist is less than that for a hole in the housing itself, a bushing made of the same material as the housing would be weaker than for the original hole.

Another reason for specifying a harder material for the bushing is based on the assumption that the conditions that caused a hole in a part already in service to wear (and thus elongate beyond an acceptable point), would be less likely to elongate a hole in a bushing made of a harder material. This does not apply to a hole in a brand new part damaged during the original drilling operation.

There are limits to the use of a bushing. One limit is the practical manufacturing limit to the thinness of a bushing's wall. Below a wall thickness of around .032 inch (which would give a bushing outside diameter .064 inch larger than the inside diameter), the risk of the part breaking into pieces during machining increases. There have been bushings made with wall thicknesses as low as .025 inch, but fingers were crossed at the time. This limits the use of a bushing to holes at least .064 inch larger than the original. This is the same size required for the next standard diameter blueprint-type fastener, since fasteners generally are sized in fractional ($1/64$ inch, $1/32$ inch, $1/16$ inch, $1/8$ inch) increments in the United States, rather than decimal increments.

One advantage of a bushing is that it usually can be made locally and machined to any dimensions required. It may be less costly than having to buy a minimum purchase of expensive oversized fasteners, with a promised delivery date six months away. Diameter control also is valuable since it is easier to machine a bushing to match an odd cleanout hole diameter than to drill and ream a hole to match a particular bushing or oversize fastener diameter. This is particularly important where the amount of hole enlargement permitted is limited and drill sizes available only in $1/64$-inch or $1/32$-inch increments.

The use of bushings in conjunction with a blueprint diameter fastener also is advantageous when an oversize hole condition does not exist to the same degree (or at all) in each separate part within a packup of two or more parts.

Oversized fasteners are oversized along their full length, but the lengths of bushings can be designed to match the thickness of any individual part or combination of parts within a multiple part packup. Since the outside diameter of a bushing can similarly be controlled, it is possible to provide several individual bushings along the shank of any single fastener. This becomes increasingly possible where a packup of parts can be disassembled.

Where cleanout hole sizes vary markedly from part to part, it often is advisable to install separate bushings, of length equal to individual part thickness, and diameter equal to the individual part final cleanout hole diameter.

As an example, installations have been made using the original .189-inch diameter screw with one bushing .062-inch long with an outside diameter of .312 inch and a second bushing .125-inch long with an outside diameter of .250 inch. The portion of the shank of the screw immediately below the head required no bushing at all, since the hole in the first part of the three-part packup was not oversized. The oversized holes in parts two and three were caused by pilot holes in those parts that were out of line with the pilot hole (undersize alignment or starter hole) in the top part, and by differing amounts.

Another reason for using separate bushings along the length of an individual fastener is to control the distribution of load from part to part. When a single, full-length bushing is used, the fastener stiffness is uniformly increased, as is the bearing strength of the enlarged hole at each part of the packup. For load distribution purposes it may be desirable to either ream a clearance hole in one part of the packup (the part you don't want to pick up any load) or to not bush that particular part. The choice usually will be forced by the accessibility of each part for drilling or reaming (a reamer is a finishing tool used after initial drilling to provide a smoother, closer-tolerance, finished hole). In several instances a single oversized hole was reamed through the entire length of a nonremovable packup and then two separate bushings with shoulders designed to control the final installed position of each bushing were manufactured and installed.

The middle of the three parts ended up with no bushing, thus not picking up any load. The shoulders on the two bushings, each installed from the opposite end of the packup, prevented them from being squeezed into the hole in the center part. A similar scheme could be used for two different-diameter bushings along the same centerline—the unshouldered length of the bushing being equal to the thickness of the part just below the screw head. This scheme also can be used where a protruding head on the bushing is acceptable or where a flush head is required. If flushness is required, the head of the bushing must be of the configuration commonly used on flush-head screws for installation into a 100° (or other angle) countersink in the surface of the part.

The ACRES Sleeve

Until recently, the minimum practical wall size for a machined bushing has required that most holes oversize by 1/32 inch or less have a 1/64 inch or 1/32 inch oversize fastener installed. In the mid-1970s, a thin-wall sleeve developed by J. O. King, Jr. of Atlanta, Georgia, and originally designed for sealing and hole protection against corrosion, came into use as a thin well bushing known as the ACRES sleeve. The acronym ACRES is taken from the first letters of its catalog description—"Anti-Corrosion capability, Repair of holes, Enlargement of fasteners, Sealing-type fastener sleeves for cost reduction in high-strength bolted and riveted assemblies."

These sleeves (now made by the Huck Manufacturing Company) are sized to fit into holes 1/64-inch and 1/32-inch oversize, thus having wall thicknesses of 1/128 inch (approximately .008 inch) and 1/64 inch (approximately .016 inch), respectively. They are made of various materials, the strongest of which reportedly is A286 stainless steel. This sleeve is not machined, but upset on special equipment using a proprietary process and having properties that require testing to determine strength allowables.

Reference to ACRES sleeves in MIL-HDBK-5 suggests some equivalency of the A286 sleeve to bolted joints without sleeves. No allowable strengths are given.

The ACRES sleeves with which the author is familiar have either flat 180° protruding heads or 100° conical flush-type heads for insertion into a 100° surface countersink. The sleeves are nominally one-inch long and have shallow grooves around the outside diameter 1/16-inch apart to both provide a series of depressions for the retention of sealant and permit the mechanical snapping off of any excess length of the sleeve using a special snap off tool available from the manufacturer. Since the strength properties of these sleeves are not easily verified, it is suggested that the A286 sleeves be restricted to use in aluminum parts only, unless fastener loads are known to be low enough in higher-strength parts to be adequately resisted if the parts were made of aluminum. The use of ACRES sleeves and standard diameter fasteners has greatly alleviated stocking problems associated with equivalently oversized repair fasteners.

Cold Working of Fastener Holes

Greater attention to the fatigue life of parts containing hole defects has brought about a search for various hole enhancement techniques. Two of the most useful techniques are *cold working* and the use of *interference fit fasteners*. Used separately or together, the cold working or interference fit fasteners increase the fatigue resistance of the part at the location of the particular hole, either at the standard (B/P) diameter or an oversized salvage or repair diameter.

For many years *cold working* of metals has been used to improve various mechanical properties of these metals. More recently, the cold working of the walls of holes has been found to increase the fatigue life of parts at these holes. This is accomplished by various means; the most common being forcing a larger diameter hardened steel pin or mandrel through the smaller hole. Diameter control is very important. Following the cold work operation, the resulting larger hole usually is reamed out an additional small amount, but not enough to markedly relieve the beneficial effects of the initial cold work.

A simplified explanation behind the benefits of cold working of holes is that the expansion outward of the metal surrounding the hole caused by forcing a larger diameter pin through the hole results in a compressive condition, or layer, in the metal surrounding the hole. This action counters the tendency of the surface to crack under a tension condition.

A surface under tension will have a tendency to crack. Although the crack enlarges at a point of high stress concentration or sharp physical discontinuity, a surface in compression tends to close up a developing crack. The compressive stresses extend approximately one hole radius distance away from the edge of the cold-worked hole in all directions.

The actual cold work is accomplished by inserting a thin-wall stainless steel sleeve with a narrow split along its length into the hole and then pulling a tapered mandrel or solid pin through the sleeve while it is kept inside the hole. After the interference diameter mandrel is backed off and removed, the sleeve also is removed and discarded and the resulting hole reamed slightly larger. Special tools and careful control of processing techniques are necessary to ensure that the cold work has been properly accomplished. There is no obvious physical or visual verification once the final reaming has been accomplished.

Just prior to this there may be an indication along the hole where the split in the cold work sleeve was positioned. In some cases, the orientation of the split in the sleeve may be specified by design engineering. Since this process results in high stresses around the hole, the concerns about the undesirability of high press-fit-type stresses in materials having a low resistance to stress corrosion come into play. As a result, cold expansion of holes in aluminum in the short transverse grain direction, where lower material ductility exists (see Chapter 4) is not recommended.

A phenomenon not often considered when evaluating the benefits of cold working holes is the presence farther outward from the hole of a tension stress field, or area necessary to internally balance the compressive stress field closer to the hole itself. The presence of these residual tensile stresses farther away from the hole requires that attention be given to situations where shy hole edge distance exists. When the free edge of the part is closer to the center of the cold-worked hole than approximately 1.75 times the diameter of the finished hole, this effect must be taken into account.

The amount of the pre-cold-work hole diameter expansion is around 4 percent for holes in aluminum and low alloy steels of 125 ksi F_{tu} and 5.5 percent for holes in titanium and low alloy steels having allowable ultimate tensile stresses of 125 to 250 ksi.

Other ways of cold working exist, including a technique called *ballizing*, where a hardened steel ball is drawn or pushed through a smaller diameter hole. Another technique uses an unsplit sleeve that is left in place.

Use of Interference Fit Fasteners

Another enhancement technique for increasing the fatigue resistance of holes is to install an *interference fit fastener*, either with or without prior cold working of the hole itself. Actually, the use of both techniques on the same hole is considered to provide the largest increase in fatigue life, although some designers would prefer to keep these techniques available for future use in case of a problem. The fact that both approaches increase the cost of the installation tends to reinforce this practice on original design. When both techniques are specified for an initial production run, there is not much room left for fatigue life improvement other than costly material or configuration changes.

There are at least five different types of interference fit fastening systems. All result in the formation of a beneficial residual stress around the hole and serve to increase the fatigue life at the hole. An additional benefit, where solid one-piece fasteners are employed, is to provide a permanent barrier against leakage of fuel, water, or any other liquid. Often, this is a prime purpose for the use of solid interference fit fasteners in fuel tanks.

The actual interference is achieved in one of two ways, the installation of a larger pin or fastener in a smaller hole (which requires a substantial insertion force) or the expansion of the shank of a free-fit pin or fastener once it is inserted into the intended hole.

Among the types of fasteners having shanks larger than the holes within which they are to be installed are threaded pins called Hi-Tigues, manufactured by the High Shear Corporation of Torrance, California, and bimetallic rivets called Cherrybucks, manufactured by the Cherry Division of Textron, Santa Ana, California.

The bimetallic designation refers to the softer metal making up the tail of the rivet which permits bucking or upsetting of the tail while allowing the use of a harder material shank. These fasteners have shanks from .002 inch minimum larger than the close tolerance holes, to .0075 inch maximum larger depending on the fastener's basic size. The larger the fastener, the more the design interference.

The types of fasteners requiring shank expansion once in the hole to achieve an interference fit are solid and blind rivets. The blind rivet shanks, consisting of an outer sleeve and an inner stem, have an enlarged portion of the inner stem pulled partway through the outer sleeve, thus creating the interference. (These rivets are made by the Huck Manufacturing Corporation of Los Angeles, California.) Expansion of the shanks of solid rivets to achieve an interference with the hole is achieved in one of two ways. Either the rivet is purposefully oversqueezed by a carefully controlled hydraulic force to obtain the necessary shank expansion, or the rivet is installed using stress wave riveters (developed by the Grumman Corporation). This device utilizes a capacitor bank discharge like a shock wave to expand the fastener shank and the surrounding hole. Both techniques are more expensive than the requirements for the more common noninterference hole, so they are generally limited to special applications and repairs.

Machined Part Repair vs. Replacement

Machined parts usually are more difficult to repair (with the exception of defective holes, slightly undersized dimensions, bad finishes, and minor surface defects) than comparably damaged sheet metal parts. Badly damaged machined parts may have to be removed and replaced because the greater bulk of metal carries a greater load than a piece of .040-inch-thick sheet metal and therefore will require a more expansive area within which to locate and install a suitable doubler or alternate load path replacement part. The need for extensive repairs to machined (or cast, or forged) parts is not required as often if the damage occurs while the part is in the detail stage, and not joined permanently to other parts. In these instances, it often is more economical to reject the part and order a replacement, especially if one is in stock. If this can be done, the follow-on order for the next group (release) of new parts would be increased by one. For this reason (among others, including inventory control), rejection and direction for part replacement should be done formally (that is, with the proper paperwork), rather than informally.

The repair of damaged machined, cast, or forged parts while installed may be the only practical course to take, but must be physically possible to accomplish. Occasionally it may be necessary to tear down the surrounding structure to permit the installation of either a sound part of identical design or a structurally improved part. On occasion the MRB engineer will have to design another machined part to accomplish the necessary repair.

The design of special repair parts as castings or forgings is seldom practical because of the long lead time and expense necessary to design and manufacture the accompanying forging dies or casting patterns and molds. One-shot repair parts generally are machined from large billets, bar or plate stock, usually available on short notice. Forgings or castings would be specified only if the same identical repair was required for a large number of assemblies, such as all those made during the previous five or ten years. Even then, the first few repair parts would be hogged out of (machined from) bar stock until the forged or cast parts were in stock and ready for installation.

The design of machined parts, as well as castings and forgings, is highly specialized and the MRB engineer should seek professional help from engineers within the organization who specialize in their design. Some companies may employ design engineers whose task is either to assist in or review and approve the design of such parts by others. If this is not possible, the MRB engineer should follow the design and processing requirements of production-line machined parts used on the same program as that having the defect.

10 Structural Analysis—Loadings and Static Strength

Importance of and Need for Structural Analysis

The capability of undertaking a structural or stress analysis of a structure exhibiting a mechanical defect is among the most important attributes an MRB engineer can possess. This applies particularly to those MRB engineers assigned to tasks in support of a production line for mechanical assemblies, as opposed to electrical components or assemblies. Recruiters for MRB engineering talent to be assigned to an operation manufacturing or purchasing detail parts or structural assemblies always are on the lookout for engineers possessing a background in stress analysis along with the ability to apply this knowledge to production floor needs.

For the sake of argument, a structural assembly can be defined as a grouping or attachment of individual parts having as a prime function the support of determinate loads and requiring an analysis of the capability of the structure to satisfactorily resist these loads during use. Another definition might be an assembly of parts, of which the failure of any one might adversely affect the purpose for which the assembly was designed. In reality, all assemblies carry loads, if only due to the weight of the parts themselves. In practice, however, the assemblies making up the skeleton of the unit usually are considered structural, whereas the smaller, more thinly made assemblies supporting miscellaneous equipment are classified as nonstructural. The distinction often is not precise and the parts of a nonstructural assembly actually supporting a piece of heavy equipment may have been highly analyzed, whereas the subsidiary parts providing additional stiffness, may have had only a cursory review. It is among the tasks of the MRB engineer to determine whether any particular part exhibiting a defect is structural or nonstructural, whether the function of the part has been unduly compromised, and whether a stress analysis must be undertaken.

Loads for Analysis

The concept of a margin of safety (MS) has been discussed in some detail in Chapter 8 and the fatigue life scatter factor (SF) in Chapters 3 and 8. The static strength of a part is a function of the load for which it is designed. This load is called the *ultimate load* and generally is the maximum load expected to occur in service, but increased by some factor such as 50 percent as specified by the customer (see Chapter 8, p. 103). The stress on the part (see Chapter 3, p. 24) is the design or ultimate load apportioned to the particular area of the part that is expected and designed to resist (or carry) the load, and is expressed in pounds (of load) per square inch of resisting area or psi. Then the design stress is compared with the allowable stress that the particular material from which the part is made can resist without failure. These allowable stresses are published in MIL-HDBK-5 and other sources, including those published by the manufacturers of the material.

The loads from which the design stresses are developed are a combination of *applied*, *aerodynamic*, and *inertia* loads.

Applied loads are those specific loads expected to be applied to the part during the operation of the vehicle (within which the part is installed) and are applied to the part from either mating parts or contact with an abutting member. A section of a longeron (a lengthwise structural member designed to carry loads parallel to its length) would experience loads from the adjoining parts of the longeron to which it is permanently fastened. A landing gear would experience the load applied to it from the ground or runway surface, both when at rest and during takeoff and on landing at varying speeds and attitudes. For example, the rudder pedal would feel the load applied by the pilot's foot as he or she pushes against it to control the direction of the airplane. Many of these loads must be calculated in advance of the design, others are mandated by contract or specification. All must be included in the analyses prepared by the stress department during the design's development.

Aerodynamic loads are those loads acting against (sometimes away from) the surface of a part exposed to the airstream through which the part moves when in operation. Both surface vehicles (such as cars and trains) and flight vehicles (such as aircraft) experience this. The nautical equivalent would be hydrodynamic loads on boats, at least below the waterline. These loads generally are established by test or analysis and are a major part of the total loading on both fixed and movable aircraft skin surfaces. They generally are distributed rather than point loads and are applied along the entire surface of a part exposed to the airstream, although they may vary greatly in magnitude from the forward end of the part to the rear end, and from outboard to inboard.

Inertia loads are those loads caused by the weight of an object experiencing an acceleration in movement during use and are directly related to the amount of acceleration experienced by the part. The inertia load experienced by a 100-pound pump accelerated to 7 g_s would be 700 pounds applied to the structure to which the pump is mounted and in the direction opposite to the direction of acceleration. For example, this pump accelerated forward to 10 g_s during the catapulting of the aircraft within which it is installed would push aftward against its mounts with a total force of 1000 pounds. This force would then be transferred to the structure to which the mounts are secured.

Both detail parts and assemblies are designed for the various combinations of loads they are expected to carry, with no one condition necessarily critical for all types of stresses. One condition (for example, a rolling pullout flight condition) may bring about the maximum tensile stress in a particular part, whereas an arrested landing condition may result in the highest shear stress. The increasing use of computers has permitted a much larger number of potential conditions to be evaluated than in the past, with a printout of the stresses associated with over 100 separate conditions for the more major structure not uncommon. In some cases, these stresses are listed in ranking order; this data is kept on file.

Modify Existing or Undertake New Analysis

When a defect occurs on either what actually is or appears to be a part of structural significance, it is necessary to determine if a stress analysis has been undertaken by the stress department and if it is available in published or unpublished form. If so, a suitable modification to the analysis should be made to determine any increase in stress (and thus reduction in the MS) resulting from the defect. If the stress analysis is not available, but a listing of the design loads is, an analysis then may be undertaken by either the MRB engineer (if within his capabilities) or by the stress department.

Load Line—Unsymmetrical Material Loss Induces Bending

The following comments relate primarily to static strength determinations, although the philosophical approaches discussed may be applicable to repair considerations for both static strength and fatigue life. Ideally, the MRB engineer will be capable of undertaking the necessary analyses. If not, the analyses must be done by the cognizant member of the stress department, but the MRB engineer should, above all, be able to determine when a stress or fatigue analysis is required.

The first consideration is whether the defect reduces either the static strength or the fatigue life of the part. To make this determination, some knowledge of what causes an increase in stress is necessary. Generally any reduction in the amount of material (metal, fiberglass, or other) will cause an increase in the local stress. This increase is due not only to the loss of working material, but often to the loss of material from a location more highly stressed than from another location within the same area of the part.

Some knowledge of load lines will be helpful here. The *load line* is the line of points along the length of a part subject to an axial load and at which the load is assumed to be centered. If the cross section of the part is perfectly symmetrical (such as a solid circle) and the part is pulled or pushed at both ends, the load line will be at the center of the circular cross section. This also is the point within the part whereby the load, when applied, causes no flexing or bending of the part. A perfectly straight part would remain straight under the action of a tension load, as it would also under the action of a pure compression load; at least a compression load low enough not to cause the member to buckle sideways.

Under these conditions, the part would experience only the axial load and sidewise bending would not be a consideration. The stress experienced would be a pure tension (tensile) stress or compression (compressive) stress and would be equal to the load divided by the total cross-sectional area of the part (or P/A).

A loss of material from the part (such as from a gouge or unintended undercut during manufacturing) would reduce the cross-sectional area of the part and thus proportionately increase the axial stress. However, for no other ill effects to occur, the loss of material must have been equally distributed around the part from the original center of the part, or at least in a symmetrical manner about this same centerline.

A gouge on one side only will be unsymmetrical and an additional type of stress, called a bending stress, will be induced. The center of resistance of the part at the area of the gouge will no longer be in line with the load line and the result will be an uneven balance, which did not exist before the defect was generated.

The difference in location between the load line and the center of resistance on the part at the location where the defect exists can be determined and this distance represents an offset of the load to the part causing a necessary bending resistance within the part. This center of resistance against bending is also called the *centroid* of the part, which for a solid circular part at least is at the same location as the center of gravity.

The offset distance times the axial load is termed a *bending moment*, expressed in inch-pound units in most instances, except where the value becomes very large when foot-pounds or foot kips may be used. The stresses generated in response to or resisting the bending moment are called *bending stresses* and are in addition to any axial tension or compression stresses.

Moment of Inertia, Axial, and Bending Stresses

The calculation of bending stresses is more involved than for axial-type stresses and requires a greater understanding of the properties of the cross section of material. For bending, one must know the center (or centroid) of the area of whatever shape that encloses the cross section. Then the distance from this centroid outward to the farthest point on the cross section in the direction of bending must be determined. One also must calculate the moment of inertia of this cross section about its centroid. The formulas necessary for these calculations can be found in many texts on strength of materials, but for illustration assume the simplest case; a rectangular cross section with its shorter dimension designated as B (for base) and its longer dimension designated as H (for height).

The axial stress for this cross section when resisting a tension load P is P/A or, since the area (A) equals the length of the base (B) times the height (H), the axial stress (in this case a tension stress) which can be designated s_t, equals P/BH pounds per square inch. The term *stress* also is often designated by the letter f and the subscript t indicates a stress producing a tension on the cross section under load. A compression stress would be indicated by the symbol s_c or f_c. The intermingling of s and f to denote stress can be confusing as can be the use of capital s (S) or capital f (F) to indicate an allowable stress and the lower case s or f to indicate an applied stress. The MRB engineer must be absolutely certain of the actual symbols in use on the project with which she is associated.

As indicated previously, the bending stresses on this cross section are caused by the bending moment (force tending to bend or rotate the section times its distance from the centroid) acting on the section. These stresses can be calculated by use of the formula $f_b = MC/I$, where f_b indicates (in a similar manner as for the tensile stress) the applied stress due to bending. The term I refers to the *moment of inertia* of the cross section about its centroidal axis 90° to the anticipated plane of bending. For the rectangular section under consideration, the moment of inertia I equals $\frac{1}{12}BH^3$ a formula engraved in the memory of most engineers interested in structures.

The term C refers to the distance from the centroid (the same point as the center of gravity for many cross sections) to the point farther away from the centroid where the bending stresses are desired to be known. Since this term, C, can vary from zero at the centroid to a maximum value of $H/2$ at the extreme outer surface of the part, it becomes apparent that the bending stress varies from O at the centroid to a maximum value at the part's far edge. Hence, the use of the plural stresses. For the maximum bending stress, the value at the edge of the part (sometimes called the *extreme* or *outer fiber*) equals

$$\frac{MC}{I} = \frac{M(H/2)}{\frac{1}{12}BH^3} = \frac{6M}{BH^2},$$

a handy formula by itself, but working it out from $\frac{MC}{I}$ seems to give a clearer picture of what is happening. In most cases, we are only interested in the maximum bending stress at the outer fiber, but we must realize that the direction of the stress reverses at the centroid. A bending stress causing a tensile pull at the extreme or outermost fiber above (north of) the centroid will cause a compressive stress of the same intensity at the outer fiber south of the centroid, as long as the cross section is symmetrical. If the cross-sectional area is not symmetrical, the C dimension to the most distant fiber on the tension side will not be the same as the C dimension to the most distant fiber on the compression side, and the tension stress due to bending will not be equal to the compression stress due to bending on the part's opposite side.

For the final determination of strength, the stresses due to the axial load, which are presumed to be equal over the entire cross section, must be added to the stresses due to bending. The maximum total stress, therefore, would exist at the outer fiber on the side where the axial stress and the bending stress were of the same type, such as an axial stress causing tension (the tensile stress) and a bending stress causing tension, or $f_t + f_{bt}$, the subscripts indicating tension and/or bending. Since the axial stress is the same on the entire cross section, whereas the bending stress varies from a maximum tensile stress to a maximum compressive stress, the minimum total stress occurs along the edge of the part on the opposite side from where the maximum stress exists, and the numerical value of the bending compressive stress is subtracted from the numerical value of the axial tensile stress. Depending on the relative magnitude of these two stresses, the minimum total stress may be negative, resulting in a net compressive stress. These extreme combined stresses must be checked against the allowable stresses for the type of material under analysis. (See Figure 10.1.)

Axial stress $= P/A = P/BH = S_t = f_t$

Moment of inertia of rectangular cross section shown $= I = BH^3/12$

Bending moment $= Px = M$ tending to stretch or lengthen the outermost fibers along the north face of the section and shorten or compress the fibers along the south face of the section

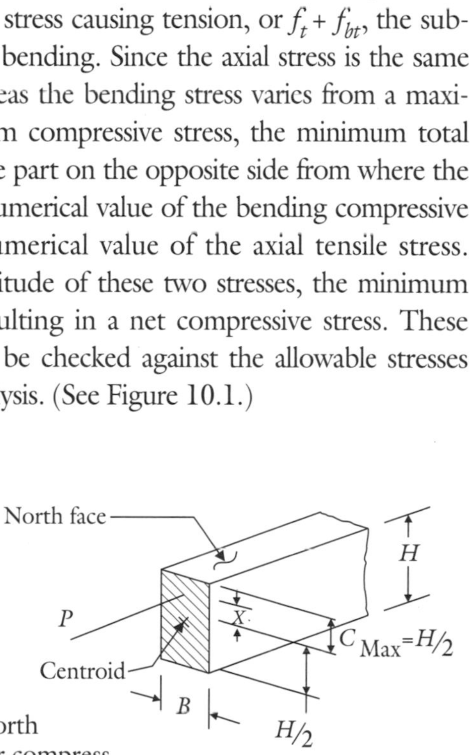

Bending stress $= MC/I = \dfrac{M(H/2)}{BH^3/12} = 6M/BH2 = f_b$

Both axial and bending stresses exist since load P is applied away from the centroid

Total stress along north face $= f_t + f_b$
Total stress along south face $= f_t - f_b$

Figure 10.1.   Stresses on rectangular cross section due to offset axial load.

As an exercise in technique it would be valuable to determine the outer fiber total stresses both for the original design undamaged cross section and the damaged tool-gouged defective cross section. This will give the analyst a feel for the effect of such damage on the strength of the part and will help build onto the store of structural judgment the MRB engineer should develop. The dimensions of the damaged part should be based on the actual shape of the part after any desirable blending off of rough edges or any other irregularities on the surface of the part at the location of the original defect. It also would be desirable for the analyst to search for any published analyses that already exist for the part under review. If such an analysis is obtainable, it often can be easily modified to show the effects of the reduced geometry on the strength of the part and also show the reviewer how the original analysis was undertaken by the design stress analyst.

Personal contact with the stress analyst is beneficial in several ways. It provides a conduit for feedback in both directions with the stress analyst being made aware of the real world activities on the production floor and the MRB engineer (if not the same person) learning of the current stress techniques and methods acceptable to the design stress group. It also may lead to the development of confidence levels between the two individuals that would not otherwise take place.

Shear Web Material Loss, Calculation of Buckling Stress

Tool gouges or other types of defects resulting in a loss of material also can occur on shear webs (see Chapter 9, pp. 115–116). This is more of a local condition and unless the penetration is quite deep will have less effect on the integrity of the web than the more common manufacture of the web to a thickness less than the blueprint minimum, either by machining or by chemical milling (see Chapter 4, pp. 60–61). Shear panels are analyzed for their ability to carry a shear load, either without buckling (a shear-resistant panel) or after buckling by diagonal tension (see Chapter 9, p. 116). The analysis for diagonal tension can be quite tedious and requires basic geometric information for not only the web itself, but for the supporting members as well as the fasteners on all four sides. Computer techniques are available. For this discussion the review will be limited to panels designated to carry the design shear without buckling.

As for tension, the basic formula for determining shear stress is load divided by area, in this case $f_s = V/A$ where f refers to applied stress, the subscript s refers to shear, V is a common symbol for the applied shear load, and A equals area (also see p. 154). (See Figure 9.1.)

When the thickness of the web is uniformly low, the shear stress is proportionately high and may result in the web being stressed beyond the point where it would otherwise resist panel buckling. To determine if this is so, it is necessary to calculate the stress at which the panel will buckle. This stress is the *buckling stress*, which is generally less than the pure shear stress the panel could carry if it were thick. There are two buckling stresses of concern. The *elastic buckling stress* is the stress that will occur within the elastic range of the material, the range of stresses whereby the web will spring back to close to its original flat configuration once the load is removed.

The *plastic buckling stress* occurs in the plastic range of the material and represents a correction to the elastic buckling stress. This text will only consider the elastic buckling stress, F_{cre1}, where the use of the capital F indicates an allowable stress as opposed to a lower case f, which generally indicates an applied stress. From Timoshenko's Theory of Elastic Stability, the elastic buckling stress $F_{cre1} = KE(t/b)^2$ where K is the buckling stress coefficient, a function of the ratio of the length of the rectangular panel's short side to the length of the panel's long side, and of the type of support the sides of the panel experience, simply supported or clamped.

If the type of support is not initially known, one can conservatively assume that all four sides are simply supported. The term E is the Modulus of Elasticity of the specific material of which the web is made, obtained from MIL-HDBK-5 or other source of allowable material strengths and parameters. E actually is the slope of the material stress–strain curve below the proportional limit of the material (more on this later). The term t equals the thickness of the web (assumed a constant thickness) and b equals the length of the web's short side. The length of the long side is called a. If the stress V/A, as applied to the web, is less than the elastic allowable web buckling stress F_{cre1}, the panel is shear resistant.

If the panel was designed to be shear resistant and the lessened web thickness causes it to buckle (according to the calculations) it can either be scrapped, reinforced with an added web to pick up all existing peripheral fasteners, allowed to go into diagonal tension (if the edge members and riveting are able to resist the extra loads induced by the diagonal tension), or a stiffening member can be added. If physically possible, the last choice often is the best because it serves to increase the panel's resistance to buckling by actually dividing it into two smaller (in other words, stiffer) panels. (See Figure 10.2.)

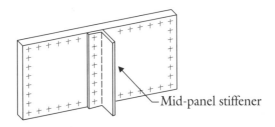

(Edge framing members not shown)

Figure 10.2.   Reinforced shear web.

Search for Current Loads, Conservative Assumptions, Equal Strength Repair

All of these considerations presume that the MRB engineer or other analyst knows the actual load that is to be applied to the shear web or panel. This also applies to axial loaded members subject to tension and/or compression loadings. The reviewing engineer is somewhat at the mercy of those individuals involved before, those responsible for any original stress or fatigue analyses that may have been required.

Among the most frustrating findings is that either no analysis was ever done (not everything is analyzed) or the analysis was not published and the original paperwork was lost. In this case, the MRB engineer should first consult the cognizant stress department authority to obtain the requisite loads and, if possible, a copy of any analysis. It also is important to determine if the loads shown are still current. A reason sometimes given by stress people for not issuing stress analyses is that the loads are expected to change as a result of oncoming tests to be conducted. The best MRB engineering analysis and disposition must be based on actual up-to-date loads.

Otherwise, a conservative assumption must be made and the result of this will be an overweight, often overcomplicated repair when, with good loads, a simple use-as-is disposition might be possible. The use of an assumed load would result in what is sometimes known as an equal strength repair: a repair that reinforces the defective structure to either equal the strength of the structure without damage or to match the maximum load that the structure can be calculated to accept from an adjacent (sometimes weaker) location on the same part.

An equal strength repair matching the strength of the original structure would be calculated to have the same applied stress as the original design. This concept is easier to talk about than to achieve, since the use of standard thickness sheet metal reinforcements, usually would result in more than the necessary thickness material rather than the precisely required thickness. Another equal strength repair would be based on the calculated ability of adjacent fasteners to pick up a maximum load (and no more) and transfer it to the part under review. The defective part might have had the ability to carry twice this load, but the repair merely has to permit it to carry the maximum possible input load. In effect, the defective area had been overdesigned and a use-as-is disposition might be acceptable if the reduced strength is no less than that required to support the lesser input load.

Load Balancing, the Free Body Diagram

A study of load transfer between parts requires some knowledge of load balancing and, a basic for all structural engineers, the free body diagram. The free body diagram is a sketch of the part under consideration, usually known as a stick diagram or single line representation, and showing all loads applied to the part for the loading condition under review. The loads are shown as arrows, the direction of the arrow indicating the load's direction, and the magnitude of the load shown as a number printed next to the arrow. Loads generally given in pounds, such as tension and compression, are represented by an arrow having a full arrowhead (both sides of the shaft).

Shear loads, when shown as a shear flow (see Chapter 9, p. 116) (that is, a running load per inch or foot along the path of application) are shown as an arrow with a half arrowhead (one side of the shaft only).

The exercise of drawing a free body diagram forces the MRB engineer to determine the point of application and the direction and magnitude of all loads, both those applied to the member from externally and those necessary to keep the member from shooting off into space (the reacting loads), sometimes called internal loads or balancing loads. The most important point of all is that for the part to remain stationary (assuming it is not meant to change its position in space), all the loads must balance in both magnitude and direction. The sum of all vertical loads must equal zero, the total of all horizontal loads must equal zero, and the tendency of the loads to rotate or twist the part about any point along the length of the part must be zero. To accomplish this, a sign convention must be set up with a plus (+) load indicating a push possibly up or to the right and a minus (−) load indicating a push down or to the left.

To account for possible rotational effects, the tendency of any load to twist the part clockwise when viewed from the front $+M$ might be considered a (+) rotational force or moment as given in inch pounds, and the reverse tendency, twisting the part counterclockwise $-M$, would be considered a negative moment. The term *moment* is equal to the magnitude of the load, generally pounds or ounces, multiplied by the distance in inches or feet from the point about which the rotational effect is desired to be known to the load itself. The actual distance used is the distance along the connecting line at a right angle (90°) to the direction of the load. The moment due to a load of 500 pounds from a point three inches directly to the side of the load line (that is, 90° away from the direction of the load) would be 3 x 500 or 1500 inch pounds, often shown as 1500"#. The moment due to the same load three inches directly in front of the load would be zero, since no rotation would be felt at this point. The ability to sketch up free body diagrams is invaluable to the engineer attempting a review and subsequent analysis of a structural member.

The free body diagram for a pure axial member would be simplicity itself. All the loads would be parallel to the long axis of the member; all would be applied at the centroid of the member; and there would be no resultant rotation or bending. All the loads would be in line and the arithmetic total of all the applied loads to the left of some point A on the part would be balanced by an internal load just to the right of point A and of the same value as the arithmetic total of the applied loads, but acting in the opposite direction.

The free body diagram of a beam, however, shows its loads generally at right angles to the long axis of the beam, thus each load tends to bend or rotate the beam.

This is normal since a beam, by definition, is meant to carry and resist loads at right angles to its length and is designed to resist this bending. The typical beam would have one or more loads applied at some point along its span or length, either as a single load at the one point or as a distributed load applied along the length of the beam uniformly or otherwise.

This distributed load would be shown as a load of a certain number of pounds per foot or inch of length if it were uniformly distributed, or as a variable number of pounds per foot or inch from one end to the other if it were not uniformly distributed along the length of the beam. For balance the beam must be held in place, generally at its ends by end loads equaling in total, but in the opposite direction to, the total of all the applied loads. These end loads are called resisting loads, reacting loads, balancing loads, or reactions.

In addition, the division of these reactions must be such that the rotational effects of the applied loads are exactly counterbalanced by the rotational resistance from the reacting or balancing loads. A simple beam (one having reacting loads at each end only) with a single applied load of 1000 pounds acting downward at its centerline will have reactions at each end of 500 pounds acting upward. Otherwise, the beam would fall downward when the centerline load was applied. If the single 1000-pound load was applied closer to the left end of the beam than to the right, the left-hand reaction would have to be larger than the right to maintain a rotational balance. The engineering term would be that the sum of the moments about the left end of the beam must equal zero, as it also must about the right end of the beam or any other point for the beam to be in balance and in state of static equilibrium. These examples are among the simplest for illustration purposes. In the real world, a beam or an axial member may experience many different loads acting together or as separate combinations of loads, but the principles of balance necessary to prepare a suitable free body diagram still apply.

Shear and Moment Diagrams

Once the free body diagram has been prepared it often is beneficial to prepare and study a shear diagram and a moment diagram, both techniques that are well covered in introductory engineering texts. As a refresher, however, they will be discussed briefly here, since both diagrams can be drawn on the same piece of paper and directly in line with or below the free body diagram. The shear diagram is a plot of the shear existing along the length of a beam or other type of structural member due to the combination of the applied and resisting loads. (See Figure 10.3.)

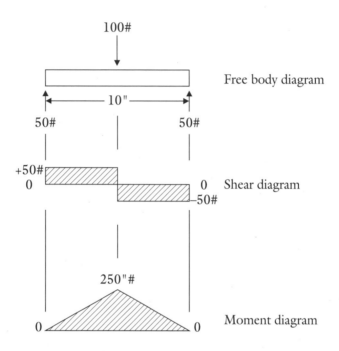

Figure 10.3. Shear and moment diagrams.

Similarly, a moment diagram is a plot of the bending moment (often shortened to the single word moment) existing along the length of the same member and due to the same loads. The shear is the load applied to the member at a particular point along the length of the member in a direction 90° to the long axis or length of the member and, if the member is in balance, should be the same magnitude whether calculated from the loads to the left of the evaluation point or from the loads to the left of the evaluation point or from the loads to the right of the same point. The shear may vary in magnitude along the length of the member depending on the type, magnitude, and direction of the applied and reacting loads.

Similarly, the bending moment can be determined at any point along the length of the member, once a sign convention has been set up by identifying, for example, a clockwise moment as a plus (+) bending moment. Once determined, the varying shear values can be plotted along an axis representing the length of the part to generate the shear diagram and the moments plotted along the same axis on the separate moment diagram. Actually, both diagrams could be plotted together, but it is more clear and understandable to plot them separately, although in line. These diagrams enable the engineer to look at the loads and visually determine where they are the greatest. With a little luck, the defect will turn out to be located in an area of lower rather than higher loads, with the result that special reinforcement may not be necessary. As a handy exercise, the engineer should construct a free body diagram, shear diagram, and moment diagram for both a simple beam supported at each end and carrying a centered load and a cantilever beam supported at only one end as if it were set into a concrete wall, but carrying the same load at the opposite free end.

Determination of Shear and Bending Stresses on an I Beam
Similar diagrams can be made for tension and compression members as an aid in determining the load at any particular point along the length of the member.

When the tension or compression member is part of an assembly (that is, permanently attached by the use of rivets, bolts, welds, or adhesives to an adjoining member, usually a shear web), the need for a free body diagram becomes even more apparent. The typical example of an axial member subject to more than end loads is the capstrip of a beam. Most people think of a beam as an I beam; so named because the cross section of such a beam is shaped like the printed capital I. When oriented in the usual manner, the middle or vertical portion of such a beam is called the *web* or *shear web* and the outermost horizontal portions (both top and bottom) are called *capstrips*.

The determination of what precise portion of the beam is called the web and what portions are called the capstrips is somewhat arbitrary for a single, one-piece beam and is much more easily defined when the beam is made up of three separate pieces fastened together.

In both cases, the area of the capstrip is considered to include a portion of the vertically oriented web, in some cases, it extends downward the same amount as the horizontal legs extend outward from the face of the web. A determination of the stresses in an I-shaped beam (actually any shaped beam) requires the use of the same formulas as for the simple rectangular beam, but with some modification. The shear stress as determined from the formula $f_s = V/A$ (see p. 147) actually is an average shear stress, but when determining the stresses on a beam, it is more correct to determine the range of shear stresses, since various areas of the beam may merit special consideration, especially where the attachments are required to secure separate parts of the beam to each other. In the case of a built-up beam made of a separate web riveted top and bottom to a T-shaped capstrip, the shear stress along the row of rivets may be critical, especially if rivets are missing or one or both members are too thin along the rivet line. In this case the shear stress is given by the formula $f_s = VQ/Ib$, which permits calculation of the shear stress at any point on the cross section.[1] The actual stress is a maximum at the center line (℄) of the I beam, if the beam is symmetrical about the (℄) and a minimum at the outermost fiber of the beam. Without getting involved in the derivation of this formula, it can be stated that the term I is the moment of inertia of the entire cross section, b is the thickness of the member at the location on the beam under review (on the web it would be the thickness of the web) and the term Q is the static moment of that portion of the beam outward of the point on the beam under consideration. The static moment actually is equal to the product of the area of that portion of the beam outward of the point of consideration times the distance of the centroid of that particular area from the (℄), or centroid of the entire beam area.

The bending stresses on our hypothetical I beam are determined by use of the same formula $f_b = MC/I$ as given on page 144, the only difference being the calculation of the moment of inertia for the entire section. This generally is done using a tabular technique given in most texts on strength of materials and is based solely on the dimensions of each element of the cross section. The values of the shear V and the bending moment M are obtained from the shear and moment diagrams.

---

[1] See reference 27 on page 240.

Use of Free Body Diagrams for Load Transfer and Load Balancing on a Built-Up I Beam

The use of free body diagrams can now be expanded further by applying them to our study of an I beam. For this purpose this text will consider that the I beam is made up of a rectangular web and two T-shaped capstrips, one along the top edge of the shear web and the other along the bottom edge. It also will be assumed that a vertical postlike member is riveted to the web's right-hand edge and serves to dump a single shear load through these rivets into the web and that a similar vertical closure member is riveted to the web's left-hand edge. Thus, there is a typical beam type of shear web, rectangular in shape and attached to framing members on all four sides.

A free body diagram of each of the separate elements of the beam, the shear web, upper and lower capstrips, and left and right vertical members (often called intercostals when between two webs) can now be sketched out freehand on a single piece of paper and to a mere approximation of scale. The key to understanding here is to position each member with a space between it and the adjoining members, a space wide enough to allow the placement of an arrow, indicating the direction of loading and of a number indicating the magnitude of the load either in pounds (if applied directly to a single point of the member, such as at its end), or in pounds per inch (or pounds per foot) if applied gradually along a specific length of the member.

Shear webs generally loaded along their edges so that a (running) load of 100 pounds per foot along the six-foot-long end of a web actually would show the per-foot distribution of a 600-pound load applied to that web and would be indicated by an arrow shaft having a half arrowhead and the number 100 printed next to it. Similarly, a single applied load of 400-pounds dumped into the right-hand end of the upper capstrip would be indicated by an arrow shaft with a full arrowhead pointing either into or away from the member depending on the direction of the load and including the number 400.

Two keys to an understanding of load distribution are the realization that the total of all loads on any one element (as represented by the free body diagram for that element) must be in balance (that is, equal to zero) if the body is to remain at rest, and that loads transfer between adjoining elements when they are in physical contact. Another necessity in understanding structures is the knowledge that parts subject to loads of whatever type experience internal stresses in attempting to resist these loads. A perfect example of this is a simple capstrip or rod subjected to a pull (tensile force) of 1000 pounds at its right end, this being the applied load, and balanced by a 1000-pound load at the left end. This load can be called a reaction, a reacting load, a resisting load, or a balancing load, depending on the terminology with which the analyst is comfortable.

By whatever name, this load is absolutely necessary to balance the member. Normally, this balancing load would be provided by the rivets securing the member to the next adjoining part further to the left. This load then becomes an applied load for this next member in line and so on throughout the entire assembly of parts. Our original capstrip under the action of this initial load will feel a stress (within the part itself) equal to 1000 pounds divided by the cross-sectional area resisting the load. The part is at rest, but nevertheless under stress and will fail if the value is more than the material can resist.

The rectangular shear web experiences a load applied parallel to and along its left edge of 600 pounds, assumed to be applied all along the 6-foot length of the edge (100 pounds per foot or 100#/ft.), and this load must be balanced by a reacting load of equal magnitude but opposite direction. The obvious place is along the opposite left edge of the web as provided by the adjoining member, so the shear web is now in balance as far as up and down motion is concerned.

Unfortunately, a look at the partially completed free body diagram will reveal that the two opposite loads, not in line with each other, will cause the web to spin like a propeller unless some counterbalancing loads are provided.

Since the shear web is connected top and bottom to the capstrips, the counterbalancing forces will be provided by the capstrips, which will then experience these added loads themselves. The value of these upper edge and lower edge shear flows is determined by calculation based on the size and shape of the web. By taking moments (that is, determining the moment due to all known loads about a corner of the four-sided web), the magnitude of the load along either the top or bottom edge can be determined and its direction must be such that its own moment is of opposite rotation to the sum of all the other moments about the same point of theoretical rotation. A little experience in doing these balancing calculations will show that for a rectangular web all four edge-shear loads or shear flows are of the same magnitude, but will differ from each other as the web geometry becomes less symmetrical and more skewed; the only final requirement being that it have four sides.

It was mentioned that the shear flow along the upper and lower edge of the web also is felt by the adjoining capstrips, but in the opposite direction.

This can be visualized by marking up the free body diagrams of the web and its adjoining capstrips to show the applied and resisting loads by use of the arrow shaft with arrowhead and the number representing the loads magnitude. Between each pair of adjoining (in reality they must be physically connected) members should be two arrows, one showing the load applied *to* one member *from* the other member and the other showing the load applied *to* the second member *from* the first member. The load designations will be the same but the arrowheads will be in opposite directions.

Now look again at the upper and lower capstrips subject to the effect of the loads from the shear web or shear panel between them. Since these shear flows are applied uniformly along the length of the web and capstrip, it becomes apparent that the capstrip will experience a buildup of load from one end to the other; from zero at one end to a maximum at the opposite end. Depending on the directions of the load, one capstrip will pick up an increasingly greater tension-type load, whereas the other will pick up an increasingly greater compression-type load. Both capstrips must be physically able to resist these changing loads along their lengths in addition to any other loads that they may also have to carry from other sources.

The ability to draw and evaluate free body diagrams and shear and moment diagrams, and to diagram load paths and load transfers is as valuable to the MRB engineer with structural responsibilities as it is to the structural designer and stress analyst.

Similar diagrams can be set up for other types of loadings (such as distributed loads and torsional (twisting) loads) and all are the basis for effective structural evaluation.

The Stress–Strain Diagram and Modulus of Elasticity

The terms stress and strain frequently are used to describe conditions of excitement and pressure felt by an individual undertaking a particularly grueling task. The choice of these words has additional meaning to the stress analyst since they are at the root of understanding the meaning of structural integrity.

Stress has been defined as the applied load divided by the area of the amount of material resisting the load, generally expressed as pounds per square inch or psi.

Strain is the amount of deflection or stretch under the effect of the load divided by the original length of the part before the load was applied, and is expressed as inches per inch. Strain can be directly measured whereas stress requires more sophisticated techniques to measure, such as the use of strain gages which are electrical devices adhesive-bonded to the part under test and measuring the change in resistance of a wire stretching along with the part.

For many materials there is a direct relationship between stress and strain and this relationship is shown on a stress-strain diagram. Most materials under modest loadings will stretch in direct proportion to the amount of load applied, at least to an upper limit of loading. Then, when the load is removed, they will contract back very close to their original position. Materials under load within this region are said to be within their elastic range. The relationship between stress and strain is linear (that is, a straight line variation); the larger the stress, the larger the strain within this elastic region. The depiction of a *stress–strain curve* or *stress–strain diagram* within this elastic region generally is accomplished by showing the increasing stress applied to a test specimen along the vertical axis of the plot or diagram and the increasing strain along the horizontal axis. Thus, the resulting curve is really a straight line sloping upward and to the right, as customarily drawn. (See Figure 10.4.)

The upper limit of this line (beyond which the material strains at a more rapid pace than below) is called the *proportional limit*, the limiting stress beyond which the strain is no longer proportional to the stress. With increasing load (and thus increasing stress), the strain increases at a more rapid rate and a plot of the stress–strain curve shows it curving more toward the horizontal.

Two additional points are identified along the extension of this curve beyond the proportional limit. The first is identified as that point along the curve at a particular level of stress called the *yield stress*. This stress is somewhat larger than the stress at the proportional limit and its actual value depends on the general shape of the stress–strain curve beyond the proportional limit. For most aluminum alloys, where both the stress and strain continue to increase with increasing load beyond the proportional limit, the yield stress is defined as that stress resulting in a specified (very small) strain such as .002 inch per inch. For all intents and purposes this is the stress level beyond which the specific material being tested will stretch at an increasing rate within the so-called plastic range of the material. Loads applied within this range will cause the material to not return to its original position after the load is removed.

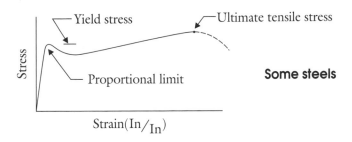

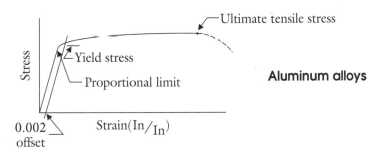

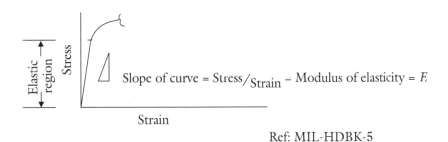

Ref: MIL-HDBK-5

Figure 10.4. Stress–strain curves.

For other types of materials (such as most steels), the stress–strain curve peaks at a stress level only slightly above the stress at the proportional limit and then reverses. The stress at this point of strain reversal is identified as the yield stress for these materials.

The second point along the stress–strain curve beyond the proportional limit is the point at the higher-yet stress called the *ultimate stress*, the point at which failure is considered to occur, at least by definition and for purposes of design. It generally is that point along the stress-strain curve where the highest stress occurs, regardless of strain, although an actual failure may not occur until a somewhat larger strain has been reached. Since a rapidly increasing strain (stretch) will prevent the specimen from adsorbing continual increases in load, the concept of an increase in strain accompanied by a decrease in load becomes understandable. The values for yield stress and ultimate stress are given in various publications of the strengths of materials such as MIL-HDBK-5, as well as various material supplier catalogs and are the values to be used for design.

The slope of the stress–strain curve within the elastic region (that is, the stress at any point along the straight line region of the curve divided by the strain at the same point) is called the *modulus of elasticity*, *E*, of that material and is expressed in psi. Generally speaking, the higher the modulus of elasticity, the stiffer the material and the greater its ability to carry loads without undue stretching or displacement, although this is only one property to consider when searching for a suitable material for a particular design or repair.

11   Structural Analysis—Fatigue Life

Follow Static Strength Check with Determination of Need for Fatigue Life Check, Seek Existing Data

The structural analyses of defective parts to determine the possibility of a resultant reduction in their longevity in service or (better stated) their fatigue life generally start with a determination of their static strength. As previously stated, this compares the value of the applied stress at the specific location of the defect (where the stress is generally larger than had the defect not occurred) with the allowable ultimate stress for the type of material under review.

If the static margin of safety is positive, the part can then be evaluated for its fatigue life in the same configuration as that for which the static strength calculations were made (that is, without repair or reinforcement). If the part is exposed in use to fatigue-type loads (that is, loads expected to be repeated many times), the reduction in calculated fatigue life may be unacceptable, although the static strength remains acceptable. If the calculations for static strength show a need for repair or reinforcement then (assuming the part is not to be scrapped) the fatigue life calculations also must be based on the physical configuration of the part with the repair/reinforcement in place. The concept of fatigue criticality and fatigue sensitivity has been previously discussed and the analyst must determine if a fatigue analysis is required, either in terms of a contract requirement, or to satisfy himself that the integrity of the repair, below the instructions for which he may be asked to affix his signature, is still satisfactory. If so, then he must either undertake such an analysis himself or have it done by someone else.

It also must be determined if the nature of the nonconformance has a negative effect on the fatigue life of the part. Some types of defects (such as too much material along the length of a part or a radius between two different thicknesses of material larger than the design maximum radius) actually may increase the fatigue life of the part if left as manufactured.

161

In general terms, any substantial increase in the stress applied to the part subjected to, and designed for, a spectrum of fatigue loadings should be a candidate for further fatigue analysis. The starting point for this analysis should be reference to any existing design fatigue analysis. If this is available, the reanalysis becomes much simpler because the fatigue loading spectrum already has been established and the S/N diagram or other fatigue life curve may still be applicable to the part under review. Since different stress analysts and engineering offices use different approaches to accomplishing fatigue analyses, it is always best to determine the precise method of analysis, stress concentration factor data, and fatigue life curves to use to analyze a particular part. All too often an analysis accomplished independently will be reviewed by the stress department and then followed with the statement, "We don't do it that way."

Background—Development of Loadings and Fatigue Curves for Determination of Service Life

In addition to the consideration of possible reductions in static strength, the MRB engineer must consider whether the nature of the defect results in a reduction of the fatigue life of the defective part or assembly. Fatigue analysis is substantially different from static stress analysis and makes greater use of statistical data. Since time spans are involved, much consideration must be given to the varying magnitudes of loads that are expected to be experienced by the assembly of parts in question during a typical span of time.

For example, a typical mission of an airplane would involve taxiing, braking, takeoff, climb, various flight maneuvers (including loads caused by wind gusts), descent, landing (at various drop or sink speeds), and braking again. For an airplane designed for use on an aircraft carrier, the takeoff would involve catapulting loads and landing would involve arrestment loads, both against various combinations of wind across the deck. Each flight would repeat these loading conditions (although the magnitudes of each load would vary somewhat from flight to flight) and the length of each flight and type of mission (such as training or combat) also would affect the magnitude and frequency of occurrence of these varying loads.

Much effort has been spent to determine a typical profile of load magnitudes and frequencies so that the effects on the fatigue life of the vehicle under study can be evaluated. For example, numerous test aircraft have been flown over many years and loading data gathered, recorded, and analyzed. Special instruments are located within the aircraft themselves to measure loads and accelerations throughout each mission. Strain gages are adhesive-bonded to various structural members and calibrated so that the amount of stretch (strain) of the underlying member under load can be measured and converted to an applied stress; the higher the strain, the higher the actual load and stress felt by the member. Accelerometers may be positioned throughout the vehicle to record the accelerations due to catapulting, flight gusts, rolling pullouts, and so forth.

The sum total of all this data is collated to show the designer the loads and accelerations to be expected in service. The order of occurrence also is important since the effects on the fatigue life of a structural part are different if the highest load is applied early during a particular time span, than if applied later during the same time span. A statistical analysis is applied to all of this data and a representative loading spectrum is developed. Mechanical test coupons are fabricated and tested to failure under the loading schedule representative of the spectrum, with the loadings applied in blocks of hours. Such tests are run continuously with the loadings applied in the proper order and magnitude and with periodic stops for crack inspection.

This description is somewhat simplified, but the end result is the development of a fatigue curve commonly called an S/N (stress versus number of cycles to failure) diagram, which shows the relationship for the particular part material and configuration represented by the test pieces, between the maximum applied stress and the anticipated number of cycles (or number of load applications) to failure. (See Figure 11.1.)

Actually the S/N curve generally is made for the application of a single repeated load and many other types of curves have been developed to allow for the application of both tension and compression loads and the varying loadings associated with a loading spectrum. In the case of spectra loadings, the vertical axis (the ordinate) of the fatigue curve shows the maximum raised stress (see p. 25) at the point of expected failure and the horizontal axis (the abcissa) shows the anticipated hours of operation to failure. (See Figure 11.2.)

The maximum raised stress at the point of failure actually is the net combined cross-sectional stress due to both bypass loads and pin loads

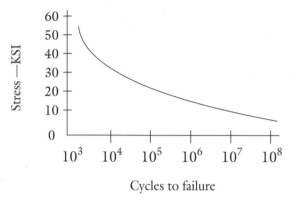

Cycles to failure

Note: Form of curve shown for illustration only.
Not to be used for calculation.

Figure 11.1.   Fatigue curve.

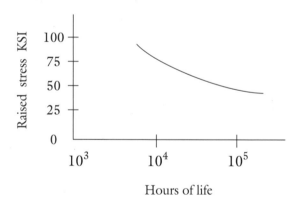

Hours of life

Ref. only: not to be used for calculation.

Figure 11.2.   Spectrum fatigue curve.

(more on this later) magnified by the appropriate stress concentration factors, K. The task of the analyst is (assuming that the appropriate fatigue
curves have been developed) to determine the actual magnified or, more
properly, raised stress at the point where the defect has its greatest negative
influence on the fatigue life of the part and then match this stress to the

curve to determine the anticipated reduced fatigue life in either cycles of repetition or hours of service.

Full-Scale Fatigue Testing, Scatter Factors Required

Completed assemblies and often the final end product itself often are required to be tested under the spectrum loadings to determine that the article has been properly designed and manufactured to survive these repeated loads without failure or (if failures do occur) to permit the design and installation of suitable reinforcements. The specification, under which the vehicle is ordered, may direct that the article be tested to twice the number of equivalent hours for which the article is designed. Since the application of loads during the test is more frequent than in actual use of the vehicle, the test program can be accomplished in months rather than years. Since the equivalent hours' worth of spectrum loadings are twice the design hours, the vehicle is tested to a scatter factor of 2.0 (see p. 26). For repair purposes, the MRB engineer may be required to design the repair to a scatter factor of 4.0, that is four times the original design life. Some details on the actual analysis required for fatigue life determination will be discussed later.

Effect of Increase in Stress on Fatigue Life

The following technique is based on the author's experience and may not represent the technique used by a particular analyst. It does, however, indicate an approach to the problem; useful in illustrating the principles involved although not necessarily state-of-the-art. Of first importance must be the realization that the fatigue life of any part, even one perfectly manufactured, will be affected not only by the basic stress level of the part at the specific location under consideration, but by highly localized areas where a specific change in the geometry of the part takes place. It is areas such as these where the basic stress is abruptly increased beyond the average stress in the part. Consider a simple rod or bar pulled on at both ends and thus under a stress equal to the magnitude of the pulling load, P (tension in this case) divided by the cross-sectional area of the bar, A. If the tension load were 5000 pounds and the part had a width of 1.0 inch and a thickness of .25 inch, the stress would equal $5000/1.0(.25)$ or 20,000 psi, a modest stress if the part were made of an aircraft-type aluminum alloy having an allowable ultimate tensile stress, F_{tu}, of perhaps as high as 76,000 psi. (See Figure 11.3.)

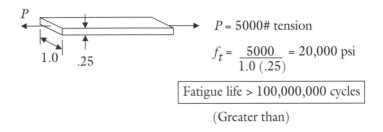

$P = 5000\#$ tension

$f_t = \dfrac{5000}{1.0\,(.25)} = 20{,}000 \text{ psi}$

Fatigue life > 100,000,000 cycles

(Greater than)

Figure 11.3.   Test bar fatigue life.

The static margin of safety would equal the allowable stress divided by the applied stress less 1.0 or $\frac{76000}{20000} - 1.0 = +2.8$ indicating an overdesign (and therefore overweight) condition since an MS of zero would also generally be acceptable. If the applied stress were 76,000 psi, the resultant MS would equal zero and this would occur if the applied load were proportionately greater than 5000 pounds, if the area of aluminum attempting to resist this load were proportionately less, or if some combination of these two conditions existed.

To determine the fatigue life of this bar it would be necessary to apply the 5000-pound load thousands of times to a test bar mounted in a testing machine and count the number of times the load was applied and released, reapplied and rereleased, until failure of the bar. Various combinations of loadings, as well as the order of the loadings, can be applied in such a machine and the results used to determine the statistical predicted fatigue life of the test bar. In this case, the load would start at zero, increase at some particular rate to 5000 pounds, and then decrease to zero again. The number of applications of this load (termed cycles of loading) to failure would be the fatigue life of the test bar under this loading.

Tests of this nature have revealed that the fatigue life is in excess of 100,000,000 cycles, for most purposes limitless, and the tests would probably have been discontinued long before failure actually occurred. However, if the width of the part had been only .33 inch instead of the original 1.0-inch width (one-third the original area since the .25-inch thickness has not changed), the applied stress would have been three times the original or 60,000 psi instead of 20,000 psi and tests of these specimens would have revealed a fatigue life of around 30,000 cycles. (See Figure 11.4.)

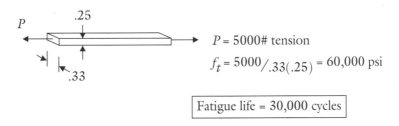

$$P = 5000\# \text{ tension}$$

$$f_t = 5000/.33(.25) = 60,000 \text{ psi}$$

Fatigue life = 30,000 cycles

Figure 11.4.   Reduced size test bar fatigue life.

Thus, the decrease in fatigue life would be far more extreme than the decrease in static strength. If the design requirement for the fatigue life were 40,000 cycles one would have to scrap (or reinforce) the part represented by the test specimen. If only 20,000 cycles were required, the part could be accepted as is, although there might be other reasons to reject it.

Stress Raiser Effect on Fatigue Life, Disproportionate Reductions Due to Added Hole and Hole Location

Unfortunately, there are other more commonly occurring changes in the geometry of a part that serve to reduce its fatigue life even further, and these are disruptions to the simple cross section just considered, a rectangle 1.0-inch wide by .25-inch thick. These disruptions to the plain, uninterrupted shape of the bar under study take many forms, some intentional and unavoidable such as holes and smooth radii between areas of differing thickness or width. Other disruptions are unintentional and hopefully, but not always, avoidable. These include defects or nonconformances such as scratches, notches, cracks, digs, gouges, or often merely enlargements of an intentional discontinuity such as an oversized hole. All of these discontinuities, intentional or not, are called stress raisers, an appropriate term since they remove material that otherwise could help carry the applied or design load. Unfortunately, their effect on the fatigue life of the part within which they occur is even greater than due merely to the reduction in area.

Tests have shown that the presence of a hole in the center of a similar part, while causing an area reduction to about 81 percent of the original area, will reduce the fatigue life to about 1.4 percent of the original (no hole in the part) life. (See Figure 11.5.)

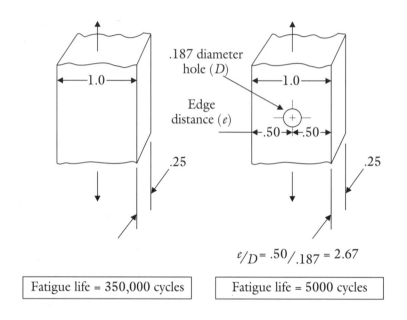

$$e/D = .50/.187 = 2.67$$

| Fatigue life = 350,000 cycles | Fatigue life = 5000 cycles |

Figure 11.5.   Fatigue life comparisons.

The reason for this is that the stress across the part with the hole in it is not uniformly constant as it is across the width of a solid part. The presence of the hole causes a redistribution of the stress so that the stress at the two edges of the hole closest to the two edges of the bar (each location 90° off the direction of the load line) increases to around 250 percent of the average cross-sectional stress. It is this raised stress, caused by the presence of the hole (called a stress raiser) that causes the dramatic decrease in the part's fatigue life. Any crack that may develop due to this peak stress generally will start at these points of maximum (raised) stress.

The position of the hole itself within the part also is extremely important in helping determine the part's fatigue life. The size or diameter of the hole in the part having a fatigue life of 1.4 percent of the no-hole fatigue life was 3/16 inch, assumed to be centered in a 1.0-inch wide part, .50 inch from each of the two edges. The edge distance of this hole (that is, the distance from the centerline of the hole to the edge of the part) is .50 inch on both sides, substantially more than the 3/16 inch (.187 inch) diameter of the hole.

The most important consideration here is the relationship between the edge distance of the hole, denoted by the letter e and the actual diameter of the hole, identified as D and both measured in the same unit of measurement such as inches.

The important number to determine is the e/D ratio, the measured edge distance divided by the hole diameter. With an edge distance in our example of .50 inch and a hole diameter of .187 inch, the e/D ratio = .50/.187 = 2.67 to both adjacent edges of the part. If the hole were located off-center, the e/D ratio to the nearer of the two edges would be less than 2.67 and to the farther of the two edges would be greater than 2.67.

In general, the greater the e/D ratio, the better, but tests have shown that with e/D ratios down to about 2.0 for metallic parts (that is, the hole centerline is a distance of two hole diameters away from the edge of the part) the reduction in fatigue life below a part with a larger e/D ratio is not critical. However, when the hole is mislocated closer to the edge or the edge itself has been partially trimmed away by accident (both conditions causing a reduction in the e/D ratio), the fatigue life decreases tremendously.

For the hypothetical 1.0-inch wide x .25-inch thick part with no hole, the fatigue life with an applied 10,000-pound load is around 350,000 cycles. With a 3/16-inch diameter hole located either at the center of the 1.0-inch wide part or no less than two-hole diameters or 3/8 inch from either edge of the part, the fatigue life of the specimen is reduced to around 5000 cycles. However, when specimens having the same size hole (3/16 inch diameter) located only 3/16 inch from either edge of the part are tested, the fatigue life drops even further, from 5000 cycles down to 1200 cycles. These reductions in fatigue life clearly reveal why unwanted holes and mislocated holes are to be avoided if at all possible and, if not, must be documented and referred to the MRB engineer for evaluation. (See Figure 11.6.)

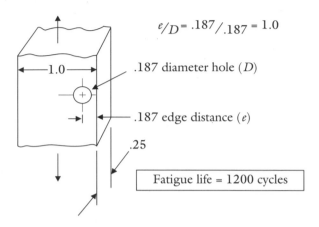

Figure 11.6.   Fatigue life with IED.

The Stress Concentration Factor *K*, Net and Raised Stress

The analysis required to determine the fatigue life of parts having such holes requires not only that the engineer know or determine the applied stress at the location of the part under consideration, but also the effects of such holes on the part's life. The goal of this investigation is to determine the amount that the stress is raised by the geometric interruption, be it a hole required by the design and properly positioned, an extra hole not called for on the blueprint, a mislocated hole, a notch cut into the edge of the part, or some other physical interruption to the smooth flow of stresses from one end of the part to the other.

Fortunately, much research has been done and data made available to assist the fatigue engineer in this task. The relationship between the raised stress and the applied stress is called the *stress concentration factor*, *K* (see p. 25). For example, when the presence of a hole in a part at a certain distance from the edge of the part results in a stress concentration factor of 3.0, an applied (average) stress of 20,000 psi at the location on the part where the hole exists actually will be boosted to 60,000 psi at the edge of the hole. The 20,000 psi would be the net stress at the hole, that is, the applied load divided by the cross-sectional area of the part reduced by the missing area due to the presence of the hole. If there were no hole, the applied stress would be called the gross stress and would be smaller because more material would be helping to carry the same load. The applied stress that the stress concentration factor boosts up usually is the net stress. Stress concentration factors generally are shown in the form of curves or plots of various types and are available for hundreds if not thousands of types of stress raisers. Certain agencies such as the Royal Aeronautical Society of England, the Aeronautical Research Institute of Sweden, the Society For Experimental Stress Analysis, and many individual authors such as R. E. Peterson,[1] have published widely in the field. The availability of this data is necessary for the undertaking of fatigue analyses for nonconforming parts.

It should be noted that the case of a single hole in a plain bar is perhaps among the most rudimentary types of stress concentration and is given for purposes of illustration only. Stress concentration factors are modified by various means and often combined with other stress concentration factors applicable to the same location to give a composite or weighted stress concentration factor and the corresponding resultant raised stress. It is not the intent of this publication to include specific formulas and techniques, but merely to outline the approach. It is highly recommended that any MRB engineer not already well versed in the techniques of fatigue analysis, spectrum loading determinations, and so forth, consult heavily with the responsible stress personnel within his or her company. This would be helpful to not only learn the particular stress techniques used, but to obtain any officially recognized load data and fatigue life curves applicable to the part under consideration.

The Pin-Loaded Hole, Edge Distance, and Material Corrections to Obtain Raised Stress and Fatigue Life

The previous example of an open hole in an axial loaded bar would apply to an extra hole never intended for the installation of a fastener. More common is a part with one or more holes intended to pick up a rivet or some sort of pin like a screw or bolt. The pin (picking up not only the part itself but one or more adjoining or mating parts) is intended to transfer a specific

---

[1] See reference 29 on page 240.

load either from or into the part under review and thus is a loaded pin. The load from this pin into the part must be determined and its magnitude will have a substantial effect on the part's fatigue life.

In structures made up of many layers of parts, a substantial effort is required to determine the distribution of loads from one part to another at each separate pin within the overall pattern of pins, whether the joints consist of only two pins or many dozen. The specific load from the pin that is applied to the individual part under review is called the *pin load*. If the pin connects (or picks up) three separate parts, the pin load for each part may be different, but the rules for a free body diagram still come into play. The sum of the three loads applied by the parts to the pin must equal zero for the pin to be in balance; otherwise it would skitter off into space.

The most common example of a pin load is that load applied by the fastener or pin at the end of a lug. In general, this load would be applied by the pin to the part in the direction acting toward the end of the lug or ear. The pin load would tend to tear out the remaining material between the pin and the end of the lug; the closer the location of the pin to the end of the lug, the weaker the lug.

In this case, the hole in which the pin is to be located has three separate edge distances. Two of the edge distances are to the opposite sides of the bar or lug; they may or may not be equal. The third edge distance is to the end (rather than the side) of the lug and is sometimes called the distance to the nearest end (in the direction of the load). All three distances are of importance in determining the fatigue life of the part. As in all instances requiring the static strength or the fatigue life of a defective part, it is necessary to determine the precise diameter of the hole, the thickness of the part in the area of the hole, the width of the part, and the distance of the center of the hole from both sides of the part as well as from the part's end. Actual measurement is best as the part may not match the drawing requirements in all instances. Once the measurements are known, the appropriate stress concentration factor curve must be obtained, the necessary dimensional parameters calculated, and the value of K read off the curve.

A material correction is often made to this value of K (often called K_T the theoretical stress concentration factor) by applying the Neuber correction.[2] This correction generally gives a somewhat lower value to the K_T based on the type of material, and leads to the calculation of the term K_n, the Neuber stress concentration factor. Once this factor is obtained, determine the raised stress by multiplying the calculated net cross-sectional stress at the hole, or the calculated bearing stress at the hole, by the corresponding K_n (for net tension or for bearing). The raised stress value can then be entered into the appropriate fatigue life curve and the life determined.

Bypass Loads, Use of Free Body Diagram to Assist in Determination

The use of two simple examples to show the progression of events necessary to determine the fatigue life of a part was purposely undertaken to lead in to the more common real-life example of a part containing more than one hole. The first example was for an open (unloaded) hole, where the applied load in the part flows past the hole just as the water in a fast-moving stream flows past an island in the center of the stream. The water has to speed up as it flows past the island, but the island itself does not add any more water to the flow. In the world of structural design and analysis, the load that flows past the hole (it does not actually flow) is called the bypass load. It is this load that is carried by the amount of remaining material in line with the hole.

The stress due to the bypass load, the bypass stress, actually is a net stress, the value of the bypass load divided by the net area at the cross section, the area of the entire cross section (as if there were no hole) minus the area taken up by the hole. No load can be carried across the hole unless it happens to be a compression load and the fastener in the hole has a tight fit.

Determination of bypass loads is relatively straightforward for a simple bar having six holes in a row with a load of 1200 pounds applied at the end hole at the right-hand end of the bar and resisted (or balanced) by a load of the same magnitude at the end hole at the left-hand end of the part.

The bypass load at either end hole is zero, since the load originates at the end hole itself, not at some point upstream of the hole's centerline location. Once this load is picked up by the part, it becomes the bypass load for the next hole in line and remains the same for each of the four center-of-part holes, unless they were to have fasteners installed and contribute some added

---

[2] See reference 31 on page 240.

loads. In this case, the two end hole loads would not be the same and the bypass loads at the four middle holes would no longer be 1200 pounds for each. The use of a free body diagram is most helpful in aiding the determination of bypass loads at each hole along a member with varying load inputs. One of the most useful ways of constructing free body diagrams is to visualize a part having the same six holes as separated into six separate but contiguous parts, each containing one of the holes at its center and each permitting the construction (that is, the sketching up on a piece of paper) of its own free body diagram.

Where holes along a common centerline exist, each free body diagram is assumed to end (or start) at a point midway between each pair of holes. Since the distance between each pair of holes in line is called the pitch or hole spacing, the free body diagrams for each hole location along a row of equally spaced (same pitch) holes would have a hypothetical length equal to one pitch. The free body diagram for each end segment or element would have one part of its length equal to one-half the pitch to the next hole and the remainder of its length equal to the distance from the center of the hole to the free end of the part, or the lug end distance. The use of this free body diagram technique is invaluable in setting up an understanding of the varying loads experienced along the length of a part subject to multiple loads along its length. The rules of balance still apply.

The determination of the bypass load can be a little tricky, but it can be readily calculated with the aid of the free body diagram showing the load at each end of the segment as well as the fastener-applied (pin) load and any shear loads applied along the shipment's length. The author starts with the larger of the two end-of-segment loads and subtracts the applied pin load from it. (See Figure 11.7.)

If a shear load exists along the length of the segment, add or subtract a portion of that shear load, depending on whether the direction of the applied shear load is the same as or opposite to the direction of the larger of the two end loads. The portion of the shear load to include would be 50 percent if the hole in the segment were halfway between the two ends, or some other percentage of the total depending on the amount the hole is away from the half-length position on the segment. Once the bypass load is known, the equivalent theoretical stress concentration factor, K_T, can be determined along with the net stress at the hole.

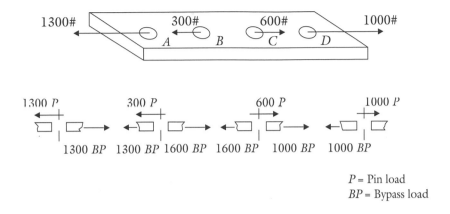

P = Pin load
BP = Bypass load

Figure 11.7.   Pin and bypass loads on free body diagrams.

Pin Load Geometry Assumptions at Other than End Hole

The effect on the fatigue life of a part subjected to a pin load at a hole not at the end of the part is undertaken in a similar manner as previously described for a pin load applied to a lug or ear at the end of the part. Since the area at the hole under consideration is not actually at the end of the part, some assumptions are required. One assumption is that the hole is considered to be in a hypothetical lug and that the distance to the end of the lug in the direction of the pin load is the same as the distance from the center of the hole to the nearest edge of the next hole in line. A second assumption is that the hypothetical lug is of a width such that the hole is centered within it. Therefore, the total width of the lug is equal to twice the lesser of the two actual edge distances from the hole's center. This permits us to determine the theoretical stress concentration factor and the net stress due to the pin load alone as before.

Use of Conservative Assumptions

The use of assumptions in these calculations may seem nonprofessional to those who are new to the field, but they serve a very necessary purpose when specific, directly applicable data, test results, and so forth, are not readily available. The analyst sets up and documents assumptions in a manner considered to be conservative, then undertakes the required analyses, using the results of these assumptions. When the missing data becomes available, the analyses can be modified to eliminate the conservatisms. What actually happens in practice is that the results of the analyses using the conservative assumptions are considered unacceptable from an economic point of view and the necessary tests are generated in an attempt to eliminate the conservatisms and show that the problem really is not a problem. The economics of the situation dictate the course of action.

Revisions to Stress Concentration Factor

Before attempting to tie together the effects of both the bypass and pin loads on the fatigue life of a part, some comments on revisions to the theoretical stress concentration factors are in order. These modifications apply only to holes. The first results in a reduction to the bypass K_T and is due to an effect called *shadowing*, a relieving effect present when holes are directly in line with each other and with the direction of the bypass loads. The edges of a hole in the shadow of the next-in-line hole experience some modest varying relief in their raised bypass stress. The amount of relief depends on the hole diameter, the edge distance of the hole, and the spacing or pitch between holes.[3]

The second modification pertains to both the pin load and the bypass load stress concentration factors, but may result in an increase in the K_{Ts} for one or both types of loadings. This effect is the result of any countersinks on the face of the hole and is basically dependent on the proportion of the length of the hole in the part in which the countersink is present as taken up by the countersink itself. For countersink depths occupying up to about 20 percent of the part thickness for pin loadings and 50 percent of the part thickness for bypass loadings, the effect of the countersink on the respective K_T is negligible, although different manufacturers may disagree. In any case, the presence of excessively deep countersinks is unwelcome and may require steps to eliminate the knife-edge that may result in the hole at the part's bottom. Additional influences on the value of fastener hole stress concentration factors due to both frictional effects and the presence of adhesive material between parts may be present, but quantitative data has not been found.

---

[3] See reference 29 on page 240.

Determination of Combined Stress Concentration Factor and Raised Stress for Hole with Both Pin and Bypass Loads

Many structural parts are loaded in such a manner that both bypass loads and pin loads are present so that some combination of the raised stress due to the two types of loading is required. Some familiarity with this is important since (in the author's experience) as many as 80 percent of all defects that affect the fatigue life of a part are due to problems with oversized, mislocated, or extra holes.

When both bypass and pin loads are present, and the part requires a fatigue life evaluation, the analyst must determine both the individual bypass load and pin load corrected stress concentration factors and the net applied stresses at the defective hole(s) in question. Then an apparent or combined effective stress concentration factor, K_{Teff}, is calculated in the form of a weighted average, dividing the sum of the bypass raised stress, $K_{TBP}(f_{nBP})$, and the pin load raised stress, $K_{TP}(f_{nP})$, by the sum of the two applied net stresses, $f_{nBP} + f_{nP}$. This K_{Teff} is then modified by the Neuber factor (p. 172) to obtain the K_N or Neuber stress concentration factor for the combined bypass and pin loadings. The resultant combined raised stress is the product of K_N times the sum of the two net stresses or $K_N(f_{nBP}+f_{nP})$. With some additional modifications, it is this combined raised stress that leads to the determination of the fatigue life from the appropriate fatigue curve.

Effect of Compression Ratio on Fatigue Life

Two modifications are the introduction of a compression factor and the use of a scale revision to match a fatigue curve to the actual condition for which the analysis was undertaken. The fatigue plot may show several curves (one each for a number of compression ratios): the ratio of the raised stress due to compression loadings to the raised stress for tension loadings. The lower this ratio, the longer the life, so a calculation must be undertaken to determine this ratio. Otherwise, assuming a ratio of 1.0, a lower calculated fatigue life would result. Determination of this ratio is based on bypass and net section pin stresses and the stress concentration factors for each and may be either positive or negative in sign: the original loads considered to be reversed in direction.

As previously mentioned, the order in which loads of varying magnitude are applied to the structure and to the fatigue test specimen has a substantial effect on the part's fatigue life. A load of such a magnitude that the stress applied to the part exceeds the tensile yield stress of the material, results in the local yielding of the part and, after release of the load, to a reduction of the strain to a value below the line of zero stress into the compressive stress region. This value, called the *compression residual stress*, then becomes the starting stress for the next application of a load and results in a comparably lower subsequent raised stress. Thus, the selection of the point in the cycle of varying loads at which this particular high load is applied can be critical to the resulting life of the part or test specimen. The use of a compression ratio with corresponding changes in the fatigue life curve is intended to account for this effect.

Raised Stress Modification Due to Fatigue Curve Scale Factor

Since a set of fatigue life curves cannot reasonably be developed for every part in a complicated structure, the test specimens used are for a particular material, stress concentration factor, loading condition, and the strain associated with the reference load for that condition, which often is the largest load in the spectrum. The actual condition for which a nearby defective part is critical may be different from the fatigue curve test specimen, so an adjustment in the raised stress for the defective part must be made. This is accounted for by multiplying the calculated raised stress at the critical point of stress concentration on the defect by the ratio of the reference strain (which is a measure of the scale of the plotted fatigue curve) to the design strain (the strain due to the maximum design condition). The fatigue curve is then entered with this adjusted raised stress to determine the life of the actual part containing the unwanted stress raiser.

12   Repair Design Considerations— Hole and Fastener Related

Highlighting of Repair Design Considerations

The design of a repair requires, as for the design of an original piece of structure, a knowledge of many things associated with the purpose of the undertaking and the means of accomplishment. The purpose of the following three chapters will be to highlight these considerations, many of which have already been stated. The design of a repair for any particular nonconformance may require knowledge of a few or of many of these subjects. They should all be available to the MRB engineer engaged in the design of repairs to mechanical structure.

Hole Diameter Callout Requirements, Selection of Diameters

The types of defects associated with bad holes have been discussed elsewhere. The importance of good holes in the repair design effort is at least as important as in the original design effort and possibly more so since the provision for a 1/32" oversized hole may be the only possible chance for salvaging a nonconforming part. For assurance that the repair mechanics drill or ream the correct size repair hole, the MRB engineer should specify the precise hole diameter required with the necessary tolerances or high and low limits for each different diameter fastener required. It is not uncommon to see a disposition stating that the special fastener should be installed in a blueprint-size hole, or in a hole sized to provide the same degree of fit as for the original. This is not necessarily incorrect, but suggests sloppy engineering. Letting (or forcing) the shop to look up the necessary size or calculate the final hole size required is not generally the best solution. The engineer should develop and supply the necessary design information. The mechanic should drill holes, buck rivets, and so forth. The need for precise hole size

179

information is even more important when an oversized fastener is required because this information may not be easily available to the shop.

As an example, consider a 3/16-inch diameter steel Hi-Lok, HL18PB-6 (made by the High Shear Corporation of Torrance, California). The nominal diameter of the unthreaded portion of the shank is .1885 to .1895 inch, a tolerance spread of only .001 inch, clearly a close tolerance fastener.

The manufacturer does not specify the hole size for the use of this fastener, but one user directs that the required hole have a diameter of .190 to .193 inch, a tolerance spread of .003 inch.

The MRB engineer should note that the minimum hole diameter of .190 inch is .0005-inch larger than the maximum specified .1895-inch diameter of the fastener (.190 minus .1895). The diameter of the unthreaded portion of the fastener shank is used because it is poor practice to use a fastener so short that its threads are against the hole in any of the parts being joined. This would be called threads-in-bearing. The sharp edges of the threads should not be counted on to carry any sideways load or be forced against the hole's smooth surface. The fact that the minimum diameter hole is to be .0005-inch larger than a maximum diameter fastener indicates that the design is for a clearance fit under all conditions of tolerance.

Now, assume that the required *B/P* hole of .190 to .193-inch diameter has been elongated in such a manner that it would take a calculated hole diameter of .215 inch on the same centerline as the original hole to clean out the hole damage.

This is nearly 1/32 inch larger, so the use of an equivalent 1/32 inch oversize same-type fastener, if available, should be considered. The appropriate repair fastener (also made by Hi-Shear) is the HL218-6 of the same length and grip as the original. The grip is the unthreaded length of the shank below the head (in the case of this protruding head fastener) and includes the head in the case of a flush head fastener. The HL218 is identified by the manufacturer as a 1/32-inch oversize fastener, although it is not quite a full 1/32-inch oversize.

To maintain the same degree of fit, the MRB engineer should compare the tolerance range of both fasteners shank diameters, calculate the tightest and loosest design fit, and then apply these fits to the diameter range of the repair fastener. If the shank diameters of both fasteners have the same range of tolerance, then a calculation can be done more quickly, but this must not be assumed. They are not always the same. Look it up to be certain, a hole once drilled cannot be shrunk (at least legally). In this case, the tightest fit would be that obtained with the maximum diameter B/P (original) fastener and the smallest legal (acceptable) B/P hole or .190 – .1895 = .0005 inch, as previously stated. The loosest fit would be with the smallest fastener and the largest allowable hole; in this case .193 – .1885 or .0045 inch. To obtain the same range of fits with the repair fastener, do the procedure in reverse. The shank diameter for the 1/32-inch oversize HL218-6 is .2172 to .2182 inch.

So, the hole size required for the tightest fit equals the largest diameter for the repair fastener plus the least clearance (slop) or .2182 + .0005 = .2187-inch diameter. Conversely, the hole size required for the loosest fit equals the smallest diameter for the repair fastener plus the largest design clearance or .2172 + .0045 = .2217-inch diameter. Thus, the repair fastener hole size required to match the range of fits from the original design = .2187 to .2217-inch diameter. This is the diameter that the MRB engineer should specify in the disposition as Ream .2187 – .2217 cleanout diameter on original hole centerline and install HL218-6 of same grip as B/P fastener.

Actually, the engineer also should specify the actual identification code for the required grip, first making sure of the correct grip required, which may not match the B/P callout because of extra shimming, and so forth. In the interest of efficiency, it would be helpful for the MRB engineering people to develop a standard listing of production type fasteners and hole sizes and oversize repair-type fasteners and their individually required hole sizes.

The preceding example was purposely selected for a clearance fit application. For an interference fit (which requires a specific interference at both ends of the tolerance range) or a transition fit (which will vary between a clearance fit if a minimum diameter fastener is used, and an interference fit if a maximum diameter fastener is used) the hole sizing technique is the same. However, care must be taken to account for the negative signs in the calculations, which should not be any problem if one realizes that with an interference fit the fastener shank diameter is larger than the hole diameter.

These calculations for the edges of the tolerance ranges are more theoretical than one might suspect because a hole is seldom at either the high or low limit of the specified diameter, nor is the shank of the fastener at its specified high or low limit. This fact sometimes can be taken advantage of by measuring the actual fastener shank diameter or the hole diameter after cleanout and may avoid unnecessary further drilling operations. For example, if a production fastener measured near or at the higher shank diameter, the fit from a slightly oversized hole might still fall within the acceptable range and the condition accepted without further work. This can only be done, however, where the fastener is to be permanently in place, not where it may be removed in service and replaced with a smaller shank (although the same type) fastener.

Interference Fit Fasteners—Use Data with Caution

The uses of interference fit fasteners to provide an increase in the fatigue life over and above the calculated (and often tested) fatigue life with a clearance fit fastener installed are becoming increasingly widespread. The technology is not particularly new, but more and more testing is under way and additional data has become available. A substantial portion of the data is proprietary, having been obtained by individual users as a necessary backup to their need to increase the design fatigue life of new products or enhance the fatigue resistance of older products or vehicles being updated to newer requirements, and/or undergoing major repair efforts.

The users of this data must be extremely careful because the bulk of it probably was generated for a particular set of circumstances to solve a unique problem affecting the fatigue life in a negative manner. Data obtained in this manner often is based on a set of design parameters that cannot be otherwise changed as they would have been if a brand new design were possible.

The manufacturers most often perform test work to solve particular problems, rather than generate the type of data that would be more in the domain of universities, fatigue specialty companies, and government testing agencies. One such company is Fatigue Technology, Inc. (150 Andover Park West, Seattle, WA 98188; telephone 206-246-2010).

Fastener manufacturers such as High Shear Corporation (2600 Skypark Drive, Torrance, CA 90509; telephone 213-755-3181), manufacturers of Hi-Tigue fasteners, also have generated much test data on the beneficial effects of interference fits. The problem in adapting this data may lie in the difference in not only the test variables, materials, and part configurations, but in the particular fatigue spectra employed.

The chances of two different fatigue tests employing the same test load spectra are remote, indeed, unless conducted in concert, or a particular user provides a fatigue loading spectrum that either the fastener manufacturer or a fatigue enhancement company uses in the development of their test data. In the absence of such commonality, the advertised benefits of interference fits and cold working should be considered more advisory than guaranteed.

Nut Torque and Fastener Shank Preload

Engineering instructions for the installation of certain threaded fasteners and their associated locknuts are sometimes accompanied by the requirement that the nut be installed with a specified wrench torque. Since this entails an extra expense, the requirement must be respected and may be mandated by contract or specification. The aim in torquing up a nut (which requires the use of a special tool called a torque wrench) is to develop an initial tensile load or stress in the fastener's shank (sometimes called the preload) and often aimed at a preload equal to about 50 percent of the fastener's tensile load capacity. This preload is related to not only the torque applied to the nut, but to the diameter of the fastener shank and the coefficient of friction between the threads on the nut and the threads on the pin. This coefficient of friction can vary greatly, depending on the surface condition of the two sets of threads (from dry and gritty to loose and greasy), so an average value of the coefficient of friction may be the best available.

The MRB engineer must realize that the actual torque value to be used for calculation purposes is the difference between the maximum applied torque wrench value and the self-locking torque (if any) for the type of nut used. Thus, if the nut has a locking feature torque of 10 inch-pounds and is cranked onto the threads of the bolt with a torque of 30 inch-pounds, the net torque tending to preload the fastener is 30 – 10 (or 20) inch-pounds. This value divided by the fastener diameter and the coefficient of friction provides the preload to the fastener.

$$\text{Fastener shank preload} = \frac{\text{Applied torque} - \text{Self-locking torque}}{\text{Coefficient of friction} \times \text{Fastener diameter}}$$

If the tolerances applicable to each of these values were to be taken into account and the extremes assumed, it would be seen that a substantial difference in the pin preload possibilities exists. These calculations are not an everyday occurrence for the MRB engineer, but serve to show a range of possibilities that the engineer may have to coexist with in both the fastener and associated repair world.

Washers—Many Types and Purposes

Washers often are considered mundane items in the hardware arsenal, but their lack of use or misuse in both design and MRB engineering repair can have serious consequences. The use of washers with repair fasteners may be more critical than otherwise, since repair constraints may impose unique requirements on the fastener system used. For example, special sealing to eliminate unusual fuel leak paths may be required and the lack of a suitably short bolt at the time needed may require the use of an available longer bolt with additional washers along the protruding unthreaded portion of the shank to prevent the nut from bottoming out on the threads before clamping against the part's face.

The MRB engineer must specify the use of the necessary additional washers, usually the same type as used elsewhere in the design.

Many special types of washers are made and the engineer should learn of their existence, configuration, and purpose. Sealing washers with elastomeric seal inserts were mentioned previously, but the MRB engineer must determine if the material from which the seal is made is the correct material for the specific fluid under consideration. A seal washer for water sealing may not be suitable for sealing a jet fuel or hydraulic oil leak.

Locking washers (both the split type and the type with serrated inner edges) are common in many products, but may be prohibited from use in the high-tech field. Some washers have a bevel or chamfer around the inside or outside diameters and are intended for use where a physical interference with the adjoining part would result. A washer with an inside diameter bevel would be used against a bolt with a large radius between its head and shank, but care must be taken to ensure that the face of the washer having the bevel or chamfer fits against the underside of the bolt head.

Otherwise, the washer edge will dig into the fastener's radius (a Murphy's Law occurrence waiting to happen). Washers with beveled outside diameters are called *radius washers* and are designed to hold the head or tail of a fastener away from a recess in the part being joined, which has its own radius close to the fastener hole's edge.

Occasionally it is necessary to fill an extra-deep countersink with a metallic filler. Rather than machine a special washer with its two faces parallel but at the same slope as the countersink, a regular flat aluminum washer can be cold dimpled to form a suitable filler and then adhesive bonded in place. For a one-shot repair, this is a cost-effective solution.

Tapered washers are very useful. They permit the use of a regular nut on a fastener installed at other than 90° to the surface of the part, rather than requiring the use of a special self-aligning nut or machining away a portion of the parent material under the nut to provide a suitable flat surface (thus weakening the part). The MRB engineer should research the availability of such washers within the organization or consider the design of one for any special repair needed. The problem with standard tapered or beveled washers is that they never seem to stock the one with the particular angle needed.

Screw Threads Described

Some knowledge of screw threads is helpful to the MRB engineer since, although he would not design a special thread form for use on a repair part, he may be presented with a nonconformance on an existing standard thread relating to one of the thread diameters. *The Screw Thread* [2] is recommended for an in-depth study of thread design.

There are many shapes and forms of threads, both straight and tapered, internal and external, right-hand and left-hand, and so forth. Many thread standards and specifications exist, and the threads themselves are grouped in various series and fit classes each having a particular purpose. The author is more familiar with the series represented by MIL-S-7742 and the newer MIL-S-8879, a higher quality thread with a controlled root radius for better fatigue resistance. For additional resistance to fatigue stresses, the external threads should be rolled during a single operation rather than machine cut, to improve the grain flow of the metal near the thread boundaries. Many parameters are involved in the manufacture of threads, from the number of imperfect threads or thread runout permitted to the desired lead-in chamfers at the ends of external threads and the entrant ends of internal threads.

Threads are identified by their major diameter, minor diameter, pitch diameter, nominal or basic diameter, number of threads per inch of length, and the thread series and class.

The major diameter is the body diameter for the threading of an external thread along a solid cylinder of material and is given with a high and low limit. The difference between the two limits is the tolerance. The sharp major diameter would be the diameter to the theoretical point of the external

---

[2] *The Screw Thread.* Bloomfield, CT: Johnson Gage Co., 1969.

thread as if it had a zero radius at the point. The minor diameter is the tap drill diameter necessary for the subsequent threading of an internal thread along the walls of a hole (the tap hole) in a piece of material and also is given with a high and low dimensional limit. The sharp minor diameter would be the diameter to the theoretical root of the external thread as if it had a zero radius at the root. The pitch diameter is midway between the sharp major diameter and the sharp minor diameter and also is given with a minimum and maximum value. The pitch is the distance between the same precise point from one thread to its adjacent thread, and the lead is the distance the threaded piece moves along the axis of travel due to one full 360 degree rotation about its axis. Pitch and lead are not necessarily identical, depending on the part's tolerances.

Another consideration for the MRB engineer is the length of thread engagement, which must be at least a minimum value for the tensile strength capability of the threaded member to be reached. A longer thread engagement will not increase the tensile strength of the joint, but a lesser thread engagement certainly will reduce it. For calculations purposes, the length of thread engagement must exclude both the usual two imperfect threads at both ends and the lead in chamfer of the internal thread. (See Figure 12.1.)

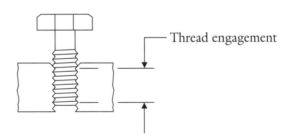

Figure 12.1.   Thread engagement.

The amount by which the thread of a bolt or screw protrudes beyond the end of a lock nut or stop nut at the end of the shank often is questioned.

If too much protrusion is evident, the bolt may be too long (also too heavy and costly) and may hit against or interfere with the operation of adjoining structure or equipment. In addition, the threads on the nut may have bottomed out along the shank of the fastener instead of clamping the parts together securely, resulting in possible movement, vibration, or fuel leaks if designed for this purpose.

The suggestion to merely cut off the excessive length of thread should be resisted until the question of acceptable clamp-up is answered. More common is the lack of adequate fastener thread protrusion beyond the nut, especially when the nut is a locking-type nut, having a special fiber or elastic insert or locally deformed threads to provide a resistance to vibrating off during use.

One thread protrusion requirement is contained in MS33588 and gives the minimum acceptable protrusion beyond the nut as a length equal to two thread pitches which may include the chamfer. Thus, a screw having a 10-32 thread (32 threads for each inch of thread length, with each thread occupying 1/32 inch) would have to protrude beyond the nut 2 x 1/32 inch or 1/16 inch. Otherwise, the effectiveness of the locking feature on the nut (and possibly the axial strength of the nut/screw threads) is questionable.

Little leeway exists for the MRB engineer to accept defective threads unless the joint is lightly stressed or overdesigned. Considerations of service use (such as threaded fastener removals required) and interchangeability must be taken into account when considering a use-as-is disposition. Tight threads sometimes can be chased or retapped, but the practice is not recommended. A tap is the cutting tool for an internal or female thread; a die is the cutting tool for an external or male thread.

Helicoil Inserts for Damaged Threads

For threads with more serious damage, a series of screw thread inserts has been developed. Originally manufactured by Heli-Coil Products (Danbury, Connecticut), they now are controlled (at least for military use) by MS33633. These inserts basically consist of a spirally wound, diamond-shaped cross section, high-strength wire that forms an internal thread configuration identical to the original when inserted into an enlarged specially threaded hole.

These inserts are used both for original design (where the hardened wire offers a greater wear and strength resistance than the softer material in which they may be installed) and for repair purposes. Both free-running and screw-locking inserts are available. Thread locking is achieved by compressing some of the coils along the length of the insert to provide a clamping action against the installed screw or bolt. Oversize inserts are used where an original part thread has been damaged and, where the damage is too severe for the satisfactory installation of these inserts, a line of twinserts is available. Twinserts consist of a pair of nested inserts, thus permitting a larger range of correction to the originally damaged threads.

Broken Tap Removal Techniques

Broken tap removal techniques are necessary when a thread tap is broken off inside a hole during the thread cutting process. The desired method of removal would utilize a pair of long-nosed pliers or an easy-out (available at hardware stores and shaped like a reverse-threaded screw). When these techniques don't work, a more sophisticated technique is required. One technique often used involves a process called *electronic discharge machining* (EDM) using a piece of equipment such as an ELOX electron drill or a spark-type tap remover. This equipment generates a high-intensity electric spark between the mandrel on the tool and the accessible end of the broken tap, slowly eroding it away until removal is complete. When this technique is used, it may be necessary to determine and evaluate the effects of the heat generated on the surrounding housing, especially if it is a heat sensitive material, such as aluminum. EDM also is used to cut close tolerance or unusual-shaped holes in metal parts.

Pin Types and Uses

Pins of various types are used in repair work, but the MRB engineer should determine if their use is permitted by the applicable program since limitations on their use may be in effect. The primary concern is that they may come loose during use and not only slip out of place, thus failing in their function, but migrate to some other location where they could jam the action of some other mechanism. To eliminate this possibility, some agencies require that such pins (which often have no shank-end threads and self-locking nuts) be permanently retained in place by some other means more positive than friction. This would be particularly applicable where safety of operation might be affected.

Many types of pins exist including the common dowel pin which may have a press fit in one or more of the parts being joined depending on whether the parts are permanently fixed in place or removable. As for all pins, the mating holes must be carefully aligned, by drilling the separate pieces with undersized holes called pilot holes and then joining the pieces together with a suitable clamping tool or aligning device and enlarging the holes to the final required diameter along the common centerline. This final operation using special tools called *core drills* or reamers may be accomplished in several steps. To aid in the installation of pins, the ends should have a light radius or chamfer to reduce the possibility of initial misalignment or dig-in.

Some dowel-type pins are tapered and may be threaded on the small end, thus ensuring a snug pull-up and assisting in their retention. Specification AN386 for taper pins and MS16555 for straight pins are among the design standards used. These are not to be confused with threaded fasteners having either flush or protruding heads and a tapered shank. Another type of pin is the roll pin (a hollow split shank spring pin) shaped in cross section like a capital C and designed to compress into a tighter C when driven into an undersize hole.

Design standard MS33547 covers the roll pin's use. These types of pins are extensively used in equipment and appliances. Another common pin is the flat-headed clevis pin with a right-angled hole drilled through the end of the unthreaded shank to accept a cotter pin or length of safety wire.

The use of such pins, generally with clearance holes, is limited to other than highly stressed or fatigue-sensitive connections.

The use of stainless steel cotter pins (see MS24665) and stainless and heat-resistant lockwire (MS9226 or MS20995) generally is governed by specifications that require specific techniques for bending and positioning the ends of each. Both devices are used to ensure that the shanks of the fasteners through which they are inserted do not slip out of the holes in which they are installed. As such, they can be called safetying members and often are directed to be installed in addition to the hexagonal nut at the threaded end of a bolt or screw. Fastener shanks can be ordered either solid or drilled for the installation of cotter pins or lockwire. When hex nuts also are used they are of the type called castellated (or castle) nuts. There are numerous slots around the nut's body to permit the passage of the cotter pin (sometimes called cotter key) or safety wire. The slots permit the castle nut to be tightened to a specific amount and then backed off (unscrewed) the minimum amount to permit the insertion of the safety device through the hole in the shank of the fastener. The resemblance of the slots around the end of these nuts to the battlements of a medieval castle tower suggests the reason for the name castle nut.

Need for Countersink Fillers

Countersink problems were previously discussed, but the subject of countersink fillers should be addressed. The need for a countersink filler becomes most apparent when an original assembly containing one or more flush fasteners with heads countersunk into a particular part is redesigned with the addition of an add-on part. This particularly applies to the installation of a new part on top of or over an existing part and intended to pick up the existing hole pattern on the original part.

When the countersink is not filled with a material at least as strong as the material in which the countersink is located, that particular part can no longer carry any fastener load except along the hole's straight shank portion. Therefore, if any appreciable load transfer is necessary as a result of the design change, it is essential to design, manufacture, and install a suitably configured filler. This filler must have an outside surface to match the slope and depth of the otherwise unwanted countersink, and an inside diameter to either match the required final hole size or left out to permit match drilling after installation. A problem with drilling a solid filler is that it may spin around the recess during the drilling operation. This usually can be prevented by applying a coating of wet paint or primer, sealant, or anaerobic retaining compound such as Loctite (see the last section of this chapter) between the countersunk surface and the filler and allowing time for set-up before drilling.

Countersink fillers sometimes can be made from the head of a flush fastener of the proper size and material by first drilling through the center of the head and then cutting away the remaining stem just below the head. Whether or not the filler might turn during enlargement of the central hole, it's a good idea to install it with wet primer or sealant to prevent moisture entrapment and reduce any possibility of galvanic corrosion if the materials are dissimilar. In cases where there is not appreciable load transfer required through holes previously countersunk, it still is advisable to fill the countersink with a sealant or a room temperature curing liquid shim material, such as an epoxy compound formulated for filling and shimming.

Bushings and Press Fit Stresses

Bushings were described previously in considerable depth. When considering the use of a bushing for repair purposes, the MRB engineer should be familiar with the type, material, size, and philosophy of bushing use throughout the original design. Some parts may be initially designed for the future use of a bushing when necessary due to normal wear. If so, this, and particularly the existence of any size limit for such bushings, would be important information for the MRB engineer. Another consideration is the need for the customer or using facility to make and install field repair bushings and the degree to which the MRB engineer at the original manufacturer's plant reduces this capability by installing a repair bushing before delivery of the part. The use of a 1/32-inch wall-thickness bushing on the original part would preempt similar use in the field, but would not rule out the use of a 1/16-inch wall-thickness bushing in the field.

On the other hand, since normal wear at a hole would be between the fastener and the inside diameter of a bushing, field repair personnel would not have to enlarge the housing hole, but merely remove the worn bushing and manufacture and install an identical replacement bushing. For this reason it would be beneficial for the MRB engineer to specify a bushing of a configuration and material readily reproducible in the field. Materials such as beryllium copper may not be as obtainable in the field as low alloy steels, and the development of an inventory of such materials in case of possible need would be costly, indeed.

In general, the engineer should configure his or her repair bushings to closely match production type bushings. An example of an MRB engineering disposition to accomplish this would be to "Manufacture and install special bushing, part number XYZ, in place of and same as production part number B65921-13 except that outside diameter should equal .776–.786 inch instead of .745–.755 inch." In all other respects the bushing size, material, heat treat, and finish would be the same as for the referenced bushing. A similar technique can accomplish a change in any of the other bushing particulars, such as inside diameter (ID), length, shoulder thickness, or material.

At this point some consideration of press fit stresses is necessary. *Press fit stresses* are tangential stresses along the inner curved surface of the housing (the part in which the bushing is installed) resulting from the installation of any bushing larger in diameter than the housing hole in which it is to be installed. The installation of such a bushing commonly is done to keep it from falling out in service and sometimes to act as a barrier against fluid leakage. The magnitude of the press fit stress depends on the stress concentration factor for the configuration, the material of both the bushing and the housing, the diameter of the hole in the housing, the wall thickness of the bushing, the wall thickness of the housing surrounding the bushing, and the amount of interference resulting from the larger bushing outside diameter and the smaller housing hole diameter.

To eliminate the possibility of stress corrosion, the press fit stress should be no more than one-half the value of the tensile yield stress in the transverse direction for the material from which the housing was made. This is a requirement from SD-24, the design specification for aircraft. Computer programs are available for determining press fit stresses, once the initial parameters are known. The ability of the program to vary the parameters at will makes the design of press fit bushings a less onerous task than when done longhand. The actual pressing into place of these bushings generally is done using a heavy tool (such as a hydraulically actuated arbor press), but sometimes a heavy hand mallet is used in a confined area. Care must be taken to ensure correct lineup of the bushing to the hole. A small bevel or chamfer at the entrant end of the bushing would assist this, but must be taken into account when bearing stresses, based on the effective length of the bushing, are calculated.

Another method used to install oversized bushings is with the use of cryogenics, such as dipping the bushing into liquid nitrogen (around −320° F). The shrinking due to the extreme cold permits the bushing to be installed in a room-temperature housing with a minimum of force. This technique often is called a *shrink fit* and is particularly useful for large diameter thin wall bushings, sometimes called *sleeves*.

The Use of Anaerobic Retaining Compound

Loctite (a proprietary compound manufactured in many versions by the Loctite Corporation of Newington, CT) is one of the family of anaerobic retaining compounds (see MIL-R-46082A) used to prevent the rotation of screws, bolts, bushings, and similar devices within their mounting holes and under the vibratory motions experienced in service. These compounds harden (cure) in the absence of air, as in a tight-fitting joint between a bushing and its housing. Loctite commonly is used as an aid to the retention of pins and bushings where not prohibited by specification and when a suitable mechanical means to retain these parts cannot be designed into the installation. For example, a shoulder bushing could not be pushed farther into its mounting hole when the shoulder is on the outside, but could possibly be pulled back out of the hole during use if not trapped by another means. The MRB engineer must give careful attention to this possibility when designing a repair bushing.

Loctite is sometimes combined with a press fit to provide further assurance against inadvertent removal. Its use requires careful cleaning and sometimes the application of a surface primer to both mating surfaces. Cure can be accomplished at either room temperature or a shorter time at an elevated temperature depending on the manufacturer's instructions. Ultraviolet light may be used to verify that the compound has been properly placed in the required joint.

13 Repair Design Considerations— Part Selection and Design

Flat Pattern Development

The ability to design and lay out the flat pattern for a sheet metal formed (or bent) repair part is a valuable attribute of the MRB engineer assigned to design repairs for defective structural assemblies. This capability may be developed during a stint in the structural design department or learned in school, but is generally not a part of the repertoire of the stress analyst. It is, however, one of the first things a structural repair design MRB engineer should learn if he is not only going to provide the necessary design information for the fabrication of new repair parts, but also evaluate possible causes leading to the misfit of existing parts.

Generally speaking, the flat pattern for a constant thickness part meant to be bent along an axis parallel to its length is that flattened-out shape of the part that (once bent to the specified angle of bend and with the required radius of bend) will result in the necessary configuration of the part. Since the path of the part around the radius of bend (the bend radius) is shorter than the path of the part if it were able to be bent to a zero radius (a physical impossibility), the flat pattern width is narrower than the sum of the projected widths of all the formed part's elements. Formed parts generally are shown both as the final shape and size desired after forming and as the flat pattern necessary before the forming operation is accomplished. This is especially necessary if the widths of the individual elements of the cross section are to be trimmed to size before forming, rather than trimmed to size from an extra wide part after forming. Trimming while flat always is easier. The MRB engineer must realize that a simple formed angle of .062-inch-thick material bent at a 90° angle around an inside radius of .250 inch and meant to have a base leg (horizontal leg) 1.0-inch wide, and an upstanding (vertical) leg or flange .875-inch wide must have a flat pattern total width of less than the sum of 1.0 and .875 inch.

The actual amount by which the upstanding flange flat pattern width is less than .875 inch is called the J value and is dependent on the thickness of the part, the bend radius and the degree of the bend, whether 90° (a right angle bend), more than 90° (a closed angle), or less than 90° (an open angle).

Charts are available for the determination of this J value which must be subtracted from the final desired width of the flange to be bent in order to determine the true flat pattern width for this flange. The total flat pattern width would be the 1.0-inch width of the base leg of the angle plus the final desired .875-inch width of the upright leg of the angle less the .189-inch J value, for a total width of 1.686 inches. This is substantially less than the 1.875-inch sum of the final formed widths of the angle's two legs. For bent-up sections containing additional flanges beyond that required for a simple angle, this exercise must be accomplished for each additional flange.

The choice of flange angles is limited to those required for the proper fit of the part, but a 90° angle probably is the most common. Closed angles can cause problems when a row of rivets close to the bend radius requires access room for the rivet (or other type of fastener) installation tool. The closed flange edge always seems to interfere with the shank of the rivet gun or bucking bar.

The bend radius chosen should be the largest possible in terms of fit because there are forming limitations on the size of this radius as the material to be formed becomes thicker or harder. For this reason, parts with tight (small) bend radius requirements may have to be formed in a softened temper condition and then heat-treated to a stronger (increased tensile strength) temper after forming. In some cases, heated forming dies may be required. The bend radius referred to is the inside radius of the bend, not the outside radius, which is larger.

Some further definitions of words applicable to flat pattern development and the engineering views of the cross sections of bent-up parts may be helpful. The definitions are applicable whether the flanges are bent up, bent down, bent to an open angle, or bent to a closed angle. For simplicity, visualize a simple angle with equal legs (the width of each leg is the same) with the upstanding leg bent up at 90°, around a generous inside bend radius. Assume that the distance between the inside face and the outside face of the part (actually the thickness of the sheet material from which the part is to be made) is constant as our visual scan of this part moves along the horizontal flange, around the curve of the bend radius, and upward along the vertical leg or flange. In practice the thickness of the actual part may decrease a little or neck down or wrinkle in the area of the radius, but this is not generally taken into account during the drawing or drafting of the angle cross section. The trick now is to identify both the inner and outer surfaces of both legs or flanges and pencil in the locations of these four faces as if there were no bend radius.

These make-believe extension lines are drawn with a light touch and will result in an intersection of the two outer face lines at a point that can be identified as the outer apex, but is officially called the point at the outside mold line (OML). The OML itself extends along the full length of the part. This actually is an imaginary line that exists in space and on the engineering drawing of the part only, but doesn't touch the part itself, unless the part could somehow be made with a zero bend radius. This could only be done with a part machined from a solid block of material, instead of bent up from a flat sheet. In a similar manner, the extension lines from the inner faces of the two legs of this angle will intersect at the point at the inside mold line (IML). This imaginary line also is often shown on engineering drawings and, depending on the part thickness, bend angle, and bend radius, may or may not lie within the confines of the part itself. A fascinating exercise is to sketch up a cross section of a 90° angle and locate with a black dot the axis of the IML and OML, then repeat the process for an angle with its upstanding flange closed 70° (from the vertical) and for another angle with its flange open 70° from the vertical. The difference in the locations of the six mold lines is astounding.

Another line to define and locate on the sample angle cross section is the bend tangent line, one for each separate leg or flange. This is the line about which the bend radius of the part starts to curve away from the flat portion of the individual flange or leg. For more complicated parts where the part may take the shape of an S, or where the radius of bend may change from a tight radius to a more generous radius, the bend tangent line is the line about which the radius change or radius direction switches from one to the other. Sometimes the bend tangent line is simply identified as the bend line (BL), but this can be confused with the buttline (p. 88). If the bend tangent lines were extended through the thickness of the part at the start and stop of the bend radius curve, the two lines from the sample angle would intersect at the axis of the inside radius (the bend radius) of the part. This would be a good way to check the validity of the exercise. A final term used in flat pattern development is the setback (SB). This is defined as the distance for each individual member or leg of a bent-up section from the OML to the bend tangent line of that particular leg.

The most correct design for a bent-up or formed part by the MRB engineer would be a full-scale layout on dimensionally stable drafting vellum with all critical sizes fully dimensioned. The minimum effort would be to describe what was desired in terms of material type, thickness, temper, and finish with possibly a statement added saying, "Hold two fastener diameters edge distance on all fasteners picked up; followed by the words "form to fit." This leaves it up to either the mechanic or the individual writing the repair procedures for the shop (or both) to do some engineering of their own. For an experienced sheet metal mechanic this may work fine, but the engineer must be assured of the mechanic's capabilities before taking this approach. There has been some excellent workmanship accomplished with these minimum instructions, but the risk is that an inexperienced mechanic will start making incorrect assumptions. For many purposes a freehand sketch fully dimensioned is better than a full-scale vellum, which may take unnecessary hours to prepare. Freehand sketches in perspective are particularly useful if the MRB engineer has the ability to churn them out. In some cases where extensive curvature is required, a full-sized layout may be necessary or the engineer must make available to the shop a full-size outline or template of the complicated area. The increasing use of computerized techniques such as CATIA or CADAM makes the development of complicated shapes more practical, but this capability must be matched by the manufacturing side of the house.

Use of Undimensioned Drawings and White Masters

Parts that are so extensively curved in outline that precise dimensioning is impractical and where precise tolerances are unnecessary often are drawn full scale on undimensioned drawings made of a dimensionally stable material such as Mylar. This applies especially to parts where the basic surface (excluding flanges, ears, and so forth) is described by a flat plane. Actually there may be many dimensions on these drawings, but the variously curved outline portions may be undimensioned and the parts made from templates matching the outlines on these drawings and then photo reproduced. In this manner the term full scale means what it says.

When parts are designed to lie along a varying curved surface, they may be shown full scale on an engineering white master, either rigid or flexible. The rigid masters are basically plaster-type forms built up to full size from appropriately contoured ribs or frames spaced a fixed distance apart and produced in the mold shop. These masters are contoured to either the outside surface of the vehicle or to a fixed distance in from the outside (or air passage) surface, generally the distance equalling the anticipated thickness of the skin. Various materials are used (other than the old-fashioned plaster) and the surface is faired (blended) smoothly to simulate the surface of the finished product. This surface then becomes the officially correct pattern for the development of further aids to manufacturing. All of the geographical reference lines and reference planes such as the stations, buttlines, and waterlines (as they intersect the surface of the white master) are permanently formed (either raised or recessed) on the surface. Then the precise location of all fastener centerlines, skin edge trim lines and cutouts, chem-mill lines, and the accompanying part identification numbers are drawn or inscribed on the surface. The finished rigid white master often is large and heavy and may require the use of mechanical equipment (such as a forklift) to move it about. In most cases the MRB or design engineer will visit the official location of the white master rather than asking that it be brought to the manufacturing floor.

Flexible white masters are thin fiberglass layups contoured to match the curves of the full-scale article and thus more portable than the rigid whites. Usually the same information is inscribed on them as on the rigid whites. As a family, the white masters, plaster mockups, and other similar full-scale rigid masters may become a vanishing breed and new MRB engineers may never see one unless they are assigned to work on older vehicles inducted into the manufacturing plant for repair, refurbishment, or modification. Design in the late 1900s increasingly is being accomplished by the use of sophisticated computer techniques such as CADAM, CATIA, and so forth. This permits the generation, storage, retrieval, and reproduction of almost unlimited geometric data in and from the computer. However, it does require that all potential users be suitably trained in the use of the necessary techniques, a requirement that is not as close to accomplishment as would be desirable. For individuals remembering the first color TV as a novelty, it may be too late. The most receptive to the current and future computerization of data necessary for the design and manufacture of actual products are those individuals who have had their own computer at home since high school or earlier.

Selection of Tolerances

The selection of suitable tolerances to be used with the design of a repair part most often matches the tolerance used for the adjoining or replaced production parts. This ensures that the shop has the facilities to obtain the same tolerances on the repair parts. A common design practice is to specify the same tolerance for most, if not all, the dimensions on an individual part. The benefits of a common or universal tolerance cannot be denied. For example, a tolerance of ±.010 inch (common for any dimensions having three decimal places such as 5.189) often is applied to machined part dimensions, except for closer tolerance holes. Similarly, a tolerance of ±.030 (common for any dimension having two decimal places such as 5.19) often is applied to sheet metal and lofted parts (those shown without dimensions on a full scale, dimensionally stable engineering drawing). A tolerance of ±.10 may be applied to a dimension having one decimal point (such as 5.2). The fact, however, that a repair part usually is one-of-a-kind requiring a special tool setup, may allow the MRB engineer to be more generous (or sometimes more restrictive) with tolerances. Usually, the tighter the tolerance, the more expensive the part and vice versa. The repair part designer also has the advantage of knowing, or being able to determine, the precise size of the parts requiring repair so she can afford to be more generous with her choice of tolerances than the original designer. This capability also may exist with the choice of angular tolerances, usually ± 1/2°, but plus or minus 5° may be quite acceptable for a specific repair part.

The Use of Tooling Holes

Tooling holes are those particular holes designed to be located on a production part in such a manner as to facilitate the manufacture of the part and/or materially aid in the placement or location of the part within the assembly of which it becomes a member. Ideally, the size and precise location of the tooling hole (as well as its identification) will be shown on the engineering drawings for both the manufacture and installation of the part. This is particularly important for machined parts. The practice of placing tooling holes in sheet metal or other nonmachined parts at the option of tool engineering cannot be recommended, especially if this is done after the final review of such parts by the stress department. Not only is the structural integrity not formally attested to, but the written record of any necessary changes in size and location may be obscure, and the configuration control of these parts may be impossible to guarantee. Tooling holes in repair parts are similarly critical, particularly when repair parts must often allow for past mislocations of mating or replaced production parts.

Tooling holes are used for both mounting and locating purposes, sometimes within the approximate center of a tooling ear, an extra appendage to a part, designed to be integral with the part, but cut off and discarded after the part is otherwise satisfactorily positioned within its assembly of parts. Tooling holes should be positioned well away from other holes and part edges, steps, and other geographical changes, and preferably in low-stressed areas. They also should be positioned well away from any bends or joggles. A minimum of three tooling holes is recommended, one each near the ends of the part and the third near the center.

Good practice also would dictate that the holes not be identical for left-hand and right-hand parts to reduce the possibility that such parts might be formed or installed while upside down or backward. The use of fastener holes as tooling holes also is done, but at some risk, especially if the mechanics are not aware of their special significance. Fastener holes used as tooling holes should be undersized to the final required fastener hole size, but if accidently enlarged would result in the part being mislocated by about one-half the amount of the enlargement if enlarged on the same centerline. In all cases, the tooling holes should be clearly identified as such, both on the engineering drawings and on any tooling drawings and associated work orders or instructions, and on the parts themselves if possible. Determination of the locations and sizes requires coordination and communication between the designer and the appropriate tool engineer, and engineering must include instructions for the sealing or plugging of all tooling holes other than those to be later enlarged for the installation of blueprint fasteners. An open tooling hole is an invitation for not only misuse, but contamination or leakage.

Joggles on Sheet Material Parts

Joggles are portions of sheet material parts that are upset or purposely displaced to provide necessary clearance for an adjoining or intersecting part over which the joggled part must lie. (See Figure 13.1.)

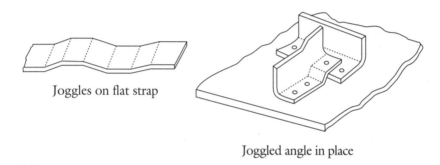

Joggles on flat strap

Joggled angle in place

Figure 13.1. Joggles.

Since the material is physically displaced by the use of a suitable joggle die, both faces of the part are relocated, with a transition or sloping ramp between the untouched flat portion of the part and the fully joggled portion. Parts of soft material may be joggled at room temperature, but harder materials must be joggled with the use of heated dies, depending on the thickness of the leg or flange to be joggled, the bend radius, and the jog's depth. Parts that are heat-treatable can be joggled while in a soft temper and then heat treated after joggling. Bent-up sections such as angles often are joggled as angles; this also requires special dies to hold the upstanding leg at the proper angle during the joggling operation. Special attention also must be given to the flat pattern and flange width that are changed in the area of the jog. Joggles generally are not done on machined, forged, or cast parts because the necessary steps can be machined, forged, or cast in place without having to reposition the material itself.

The MRB engineer may have to include a joggle in the design of a repair part and thus also have to know of the capability to produce such a part. As for other design considerations, the existence of joggles in nearby structure should be an initial tipoff to the manufacturing capabilities available. If the slightest doubt exists, the engineer should consult with tool engineering personnel associated with the particular program involved.

When specifying the use of a joggled part, the MRB engineer must be assured that no existing fasteners to be picked up or newly added repair fasteners are located so as to be installed within the sloping transition area of the joggle. If this cannot be avoided, a triangular-shaped filler piece (shim) must be designed and installed in the cavity or space between the two levels of the joggle. This is done to prevent the fastener from squeezing the joggle transition toward the mating part and possibly bringing about a premature structural failure.

The condition where a space exists between two parts meant to lie against each other, and a fastener installed in this area has a portion of its shank exposed, is called *shanking* and should be avoided. Where joggles are required in highly stressed areas, the MRB engineer should consider undertaking or having undertaken a stress analysis of the area in question since the existence of the joggle results in a load line relocation or eccentricity and may generate undesirable local bending stresses in the joggled part, especially where the joggled element is not part of an angle section, but merely a double-bent strap.

Lightening Holes Increase Stiffness

Lightening holes frequently are seen in the middle of large, flat, sheet metal webs. They are lightening holes because the circular- or otherwise-shaped cutout removes weight from the panel. They also may be used as an opening to permit the passage of equipment lines, tubing, electrical conduits, bundles of wiring, and so forth. What distinguishes these holes from others, however, is the presence of a short flange all around the hole and formed from the basic web. This flange provides the basic (often only) purpose for the lightening hole, that of increasing the web's stiffness and resistance to buckling.

Tests have shown that a given-size, thin sheet metal web with a substantial flanged lightening hole near its center will carry a higher shear stress before buckling than the same size web without the lightening hole, and simultaneously will be lighter. The presence of the flange increases the webs resistance to initial buckling, as would the addition of an angle across the web's center from one side to the other and then riveted in place. The MRB engineer should take advantage of this technique when considering a shear web replacement but, as in the case of other sheet-metal-forming techniques, be assured that the manufacturing capability exists to fabricate this type of web.

In one instance, a panel from a previously scrapped assembly which contained a suitably sized lightening hole shear web was cut out and installed at another location where such a web was beneficial. The replacement web section was attached to the remaining portion of the original web with the same type and size rivets all around as had attached the web to the remaining framing structure.

Use of Existing Extrusions and Rolled Sections

An extrusion is the particular configuration of a material that results from a fabrication process that forces the material in an initial softened stage through a series of openings to form a specific shape maintained along the full length of the piece. The material often is a metal, but the term also is applied to parts made of plastic, elastomerics (rubberlike), and so forth. The cross section of the extrusion is the same along its entire length, thus it can be cut into short pieces to easily provide a large number of identical parts.

The cost generally is less than fabricating the parts individually, especially when large quantities are needed. Extrusions also are used for long members that do not have to be tapered along their length to save weight. Architectural use is widespread. For the MRB engineer, the use of an extrusion for a repair part would be practical only if the particular shape needed were already in stock and this ready availability were verified before calling it out in a repair disposition. Many manufacturers catalog the available extrusions carried in their stock of materials. The development of a new extrusion for repair purposes generally would be prohibitive because of the high cost for the necessary tools and dies required by the fabricator.

Another form of material available in long, constant cross section lengths is the rolled section. Generally, these are made from lengths of sheet that are passed through a series of rollers (rolls) to form a final section of bent up (and/or bent down) elements, the most common of which is an angle.

Numerous shapes are available but, as in the case of extrusions, the MRB engineer must be assured that the particular section desired is carried in stock. One way to do this is to borrow a section used in the fabrication of current parts in production. Although this may be no guarantee of ready availability, it is often the starting point in the design of repair parts. One common procedure is to design a particular part, stating that it is to be made from extrusion XYZ, or rolled section ABC, or equally common, from an existing production part LMN. The MRB engineer must pay particular care to the configuration of the parent part so that the repair part can be made from the parent with a minimum of additional operations. Cutting off 1 inch of a 4-inch-long parent part is far superior to requiring that the thickness of one leg of the 4-inch-long part be reduced from the original 1/4 inch to 1/8 inch along 2 3/4 inches of its length. The less rework the better, as long as weight increases are kept to a minimum, the stress and fatigue life levels are not adversely affected, and the repair parts fit properly without the need for force or excessive shimming. Extrusions and rolled sections come in many materials, tempers, and configurations. An advantage of extrusions is the greater multiplicity of shapes possible and the fact that the square corners at the intersections of the separate legs provide more inherent stiffness than the radiused constant thickness wraparound corners of rolled sections.

Shims Fill Undesirable Gaps

Shims have been referred to several times in the preceding text and are basically fillers sized to match an unwanted cavity between poorly mating parts.

Shims often are of a standard configuration and material developed by a manufacturer of assemblies for purposes of standardization and cost reduction. The standards generally are set up to permit substantial choices in the thickness, width, and length, and are primarily intended for production rather than repair use. Other types of shims may be set up for a tapering thickness (a common requirement in the repair world), or as a peelable laminate, varying plies of which may be peeled away to obtain the final desired thickness.

The MRB engineer should become familiar with the choices of standard shims available within the company or service unit, but also should realize that, even though they are in the standards manual or catalog, they may not be in the stockroom.

Another type of shim is used when a fastener mislocation causes the fastener to fall so close to a radius of the part in which it is mislocated that the head or tail of the fastener lies against the curve of the radius rather than against the flat surface of the part against which it is supposed to lie.

Of course, the same result will occur if the hole is properly located, but the radius is mislocated. In either case the situation often is unacceptable and the MRB engineer will call for the use of a radius block, a solid block of material with one edge curved to match the radius of the basic part and with a thickness sufficient to permit the head or tail of a longer blueprint type replacement fastener to lie flat against the flat surface of the block.

Occasionally radius blocks are used in design to spread out a concentrated tension load from an individual fastener that must be located excessively far away from a nearby load balancing member.

Hinge Centerline Locations Critical

Hinges often are custom designed to fulfill a particular space requirement and when damaged must be either replaced or, if only hole damage has occurred, have a hole repair specified and accomplished. Other than strength considerations the most important consideration is that the precise location of the hinge centerline not be changed, either by the installation of a replacement hinge or repair of the damaged hole. A shift of either the hinge hole itself or the hinge mounting or attachment holes may render the repair unacceptable because the unit may not fit the assembly into which it is to be installed or permit an acceptable fit by a mating part. Hinge points often are interchangeable so that the parts having them can be installed on any assembly requiring the part.

Extreme care must be specified by the engineer and taken by tooling and manufacturing people to ensure that the replacement hinge fitting or repaired hinge point hole is located precisely where it belongs. A special tool may be required or the part may have to be rebuilt in its installed location using an undamaged mating part as a surrogate for the otherwise necessary tool or locating fixture.

Some designs use piano-type hinges and not much leeway is permitted in repairing these although a successful repair has been accomplished by removing a short, damaged section of a longer hinge and replacing it with an undamaged equally short length. In this instance, a replacement hinge pin, the same length as the original, was used to ensure the correct position of the replacement hinge section. The same method used to retain the original hinge pin could not be accomplished, so an alternate retention method had to be devised.

Material and Thickness Choices

The selection of materials and thicknesses (gages) by the MRB engineer generally follows the structural and fit requirements of the repair. Most often a review of the existing design is the starting point and repair materials and gages comparable to the original are specified. Many manufacturers publish listings of standard and/or available stock, including not only raw materials in various forms but fasteners, adhesives, and so forth. The MRB engineer should take advantage of these lists at the beginning of repair design considerations. It is most embarrassing to have a manufacturing representative inform you that "we don't stock that size sheet and it will take seven months to obtain it." Changes to the repair instructions will be required and good engineering, in the context of the material review effort, also should be attainable engineering.

14 Repair Design Considerations— Process Technology

Adhesive Purpose and Properties Must Be Known

Adhesives (sometimes called resins) are used to attach parts to each other without the need for mechanical fasteners. They frequently are used to attach nonmetallic materials (such as rubber seals and gaskets) to metal parts in order to provide a flexible barrier against leakage, to cushion parts against shock loadings, and to make up for varying gaps between mating parts. When used in this manner, the defective parts are mechanically removed and replacement parts installed with new adhesive. This seldom requires written direction by the MRB engineer because the structures can be reworked to the original blueprint configuration, the defect being eliminated at the shop level.

The use of adhesive for structural purposes, however (as in the manufacture of honeycomb panels), often requires the disposition of defects by an MRB engineer because the original structure may be damaged during the initial teardown operations or the original adhesive bonding process cannot be fully duplicated. The original honeycomb bonding process may have been done in a special tool or fixture called a bondform (shaped to nest or support the bondment only), not any additional parts that may have been permanently attached to the bondment during a latter stage of manufacture.

The repair for such a panel requires the removal of the defective parts and any debris resulting from such removal, as well as preparation of suitably sized replacement pieces of the same or stronger material. Provisions must be made for splicing these replacement pieces to the remaining original structure with the application of a suitable and compatible adhesive cured in such a fashion as to achieve its maximum strength without reducing the strength of the original materials. Specific knowledge of the properties of and the application requirements for such adhesives is absolutely necessary. The MRB engineer, if not already knowledgeable, must consult with a materials specialist. A three-person team would be ideal for this: the MRB engineer, a materials or chemical process engineer, and a nondestructive inspection specialist to test the repair area after completion to ensure that the repair was satisfactorily accomplished.

Replacement of Honeycomb Panel Damaged and Contaminated Areas

The types of defects that honeycomb panels are subject to have been extensively discussed in Chapter 5, as has the need for knowledge of the requirements for satisfactory use of the adhesives used in the manufacture of such panels. Since the structural integrity of honeycomb panels requires the satisfactory adherence of the repair adhesives to both the remaining original structure and to the newly added structural parts, it is necessary to ensure that all contamination be removed before any new adhesive is placed in position or laid up. This requires that all corrosion and corrosive by-products and all defective core be completely removed, regardless of the degree or depth of removal required. There is little point in removing half of the pitted surface of a corroded skin when a satisfactory rebound requires a sound, hospital-clean bonding surface. Both the MRB engineer and the manufacturing people must recognize this need, and the possibilities of exceeding the structural limits of the panel must be evaluated in advance of the actual repair activities. In many cases it will be obvious that the removal of corrosion from the surface of a skin will result in only a superficial (or otherwise tolerable) loss of material. In other cases, however, where surface corrosion is extreme, the MRB engineer should be consulted before the corrosion removal blend-out operations begin. Better to recognize the risk of obtaining an unrepairable panel before corrosion removal begins than after it has been completed.

The removal of all corroded core also must be complete. This requires the removal of one of the two skins to which the core is bonded. The only exception to this is when a section of removed core can be replaced through a cut out portion of the web of the closure member and this can be accomplished only when the area of required core replacement abuts an edge member. This type of core replacement can best be accomplished by cutting and installing two wedge-shaped pieces of core, so that the pieces can be forced inward and the slope of the interface between the two pieces will allow a slippage and a resulting pressure against the new bond surfaces at both the interface between the new core pieces and the interior faces at the original remaining core (as well as the existing inner and outer skins).

When replacement core is installed through an access hole in one of the face skins, it must be carefully sized to provide a snug fit all around its periphery and permit the necessary bond pressure to be applied by a mechanical weight or force against the exposed end of the core piece, which must be purposely left high. This bonding pressure is necessary to cause an imprint against the new adhesive film and ensure that each cell of the new core section is wetted with the adhesive during the cure cycle. Otherwise, an unwanted void may result.

Another requirement to ensure the satisfactory bond line integrity of the left-in-place core is that all contamination be removed. In the case of water ingestion due to a past puncture of the honeycomb bondment, the water must be completely removed by one means or another. This can be accomplished by drilling numerous water removal holes through the panel skin and/or closure member web, then installing a vacuum bag above these vent holes and subjecting the entire assembly to an extended period of heat. The higher the heat, the shorter the time period required. Too high a temperature, however, not only risks softening the existing panel parts, adhesives, sealants, and finishes, but may create the formation of steam and the resultant pressures from moisture trapped inside the panel. Such pressures have been known to extensively damage the honeycomb panel (blow the core, see p. 79) so any temperature in excess of the boiling point of water should be avoided.

The assumption in these water-removal exercises is that the trapped water is relatively clean and will not cause the surfaces against which it was in contact to become chemically contaminated. If such a suspicion exists, the only recourse is to flush the contaminated surfaces with a suitable solvent. This is easier said than done. If contamination is a reasonable possibility or the moisture within the panel is other than clean water (such as fuel, hydraulic oil, cleaning solution, and paint strippers), then the contaminated core must be removed and a thorough flushing with a suitable solvent undertaken.

The application of repair adhesives to a contaminated surface is possible. The normal curing procedures would probably result in the necessary hardening of the adhesive, but it could not be guaranteed to stick to the adjoining surfaces within the required specification bondline strengths. The possible pitfall here is that, with the adhesive physically in place, most nondestructive tests would not reveal a nonconformance and the weakened panel would move along its assigned path to eventual service. The only test that would reveal a substandard bond would be one against a similarly contaminated surface, or one against the actual repaired part itself. The trick would be in knowing which precise test location was or was not contaminated. The joking reminder from a neophyte MRB engineer, "When in doubt, throw it out," might have some credence if a safety-of-use question arises.

Core replacement is accomplished by substituting a repair core of equal or greater strength, although the greater-strength core will increase the weight, if of the same strength foil material, and also may increase the panel's local stiffness. This is not always desirable. Repairs also have been accomplished by injecting a mass of adhesive into an area of defective core, but this is not the most acceptable technique, especially for the long term. In general, the use of a repair core piece of not only the same strength, but the same or lesser cell size, is preferred since the bond line strength is affected by the merging of the cell walls into the adjoining adhesive layer. The smaller the cell size, the more load-carrying cell walls per square inch of bond surface.

Once the replacement core is in place, it becomes necessary to replace the missing skin area originally removed, either to permit the installation of the new core or to eliminate an area of skin surface corrosion. Since the choice of an adhesive-bonded honeycomb panel was originally made to permit the thin skin weight savings possible with such a panel, the very thinness of these skins (except in the area of attachment to major fittings, heavy edge closures, and so forth) will preclude the use of fasteners for skin splicing. Such splicing can be accomplished by the use of metal-to-metal, fiberglass-to-fiberglass, or other surface-to-surface adhesive bonding. The load transfer required would be based on the width of the cutout 90° to the direction of the load and the skin design stress.

If the actual load is not known, a conservative approach would be to assume that the skin is working at full capacity, that is, at its ultimate tensile stress. Therefore, a skin cutout 1-inch wide, with a thickness of .040 inch, and an allowable ultimate tensile stress of 62,000 psi could have carried a maximum load of 62,000(1.0)(.040) = 2480 pounds across this 1 inch cutout.

An adhesive bond joint would have to be long enough to carry this 2480 pounds. If the allowable adhesive lap shear stress for the bonding system used were 1500 psi, the overlap distance or length between the sound original skin surface and the repair skin would have to be 2480/1500 = 1.65 inches for each 1 inch of original skin cutout width.

If the actual skin load across this cutout were only one-half the assumed 2480-pound load, the adhesive surface skin-to-skin overlap length would only have to be 1.65/2 or .825 inch for each inch width of the cutout. The repair splice skin pieces normally would be of the same or perhaps one standard gage thicker and of the same material and temper as the original skin.

Some MRB engineers would go one step farther and add a row of small-diameter flush head, blind, tack rivets along both the edge of the cutout and the edge of the replacement skin patch to reduce the possibility of the patch peeling away from the parent skin. The peeling resistance along adhesive-bonded joints under a lap shear strength tensile loading (that is, the skins are in tension, but the adhesive provides a shearing type resistance to this tension) is not extreme. When a skin cutout is necessary to permit the underlying honeycomb core replacement, the final core insert is cut off flush with the original skin outer surface after the core insert adhesive is fully cured. The skin patch is bonded to this new surface along its entire surface area during a second bond cycle.

The Uses of Shot Peening

Shot peening is an enhancement process similar to sand blasting, but primarily used to either form a desired curvature along the surface of a relatively thin-walled part or increase the resistance of a part to the generation of fatigue cracks. Shot peening used to form parts is a sophisticated version of an exercise common in elementary school where the student in a shop or mechanical arts class is taught to form a copper dish or curved tray by continually tapping the surface of a flat piece of sheet copper or other soft metal with the ball end of a ball peen hammer head until the desired curve is attained. Shot peening accomplishes the same result by a more controlled means whereby a continual stream of steel (or other material) shot, (like BBs or small ball bearing balls) is blown through a special nozzle and against the surface to be curved. The type and size of the shot particles, the intensity of the blast, and the exposure time can be varied to provide the desired result. Some experimentation may be necessary, but (once worked out) its repeatability is excellent although some operator finesse is helpful.

This technique also can be used to apply a special satinlike or pebbly-appearing surface finish to the part being peened and also can be used to remove paints or other coatings, in which case the process is called *blasting* or *dry honing*. The choice of shot type, size, and the intensity of the stream affects the results, with shot varying from rejected ball bearing balls to glass beads, garnet particles, and walnut shells among others.

An increasingly common use of shot peening is to improve the fatigue life of metal parts. Among the earliest of such uses was the shot peening of automobile springs. The technique is to shot peen the surface of the part to a particular intensity with the designated type of shot for the necessary time to form a compressive layer along the surface. The necessary intensity of the peening must be determined by the generation of test specimens peened with varying types of shot, blast pressures, and exposure time, and then fatigue-tested to the required loading spectrum. The intensity of the peening is designated as the curved height of a standard peening test specimen called an Almen specimen (ref. MIL-S-13165B-2 and SAE Publication AMS 2430J). This is the arc height of the standard specimen after peening and is measured to the top of the specimen's curve in thousandths of an inch.

The specimen is flat before peening and typical peening intensities vary from .006 to .015 inch. The parts to be peened must be clean and bare and any areas not to be peened must be covered or masked off so that the shot will not directly strike these surfaces. Sharp corners in the path of the shot must be removed to prevent surface rollover and the used shot must be completely removed. It may be processed for possible reuse. The parts to be peened may be initially bent (with proper authorization) and Almen specimens may be required during the peening of one or a group of production parts.

Saturation shot peening results in full peening coverage over the surface of the part. Peening beyond this point may not appreciably increase the arc height of a test specimen. After peening is completed, care must be taken to limit the exposure of the peened part to high temperatures. The limiting temperature depends on the part's material—lower for aluminum, higher for steel and titanium. Careful records should be maintained for all peening operations and the operators must generally be trained and certified.

Air Hammer Peening

Air hammer peening is a technique similar to shot peening used to either form a metal part to a desired specific curvature or, conversely, flatten out a bent part meant to be straight. Both are cold working techniques. Shot peening employs a continuous blast of particles such as steel shot forced under pressure through a nozzle aimed at the surface meant to be reformed. Air hammer peening is accomplished with an impact tool such as a rivet gun fitted with a curve-ended rod held against and maneuvered along the surface being formed. The advantage of air hammer peening is that it doesn't require the special bulky equipment that shot peening does or the tentlike surrounding necessary to catch the rebounding shot. The use of both techniques requires special training, controls, and prior testing.

Crack Removal and Verification

Cracks have been discussed from the standpoint of their negative effects and the importance of acting on the suspicion of the possible presence of cracks (see pp. 43–44), but the techniques of removal are worth further discussion. The standard technique associated with the discovery of cracks has been to *stop hole* a crack at every end, generally both ends in the case of a simple single line crack, but sometimes three or more ends for a spider-web-type crack. Standard thinking was that drilling a hole at the end of a crack eliminated the extreme stress-raising effect of the point of the crack, replacing it with a smooth, round hole that would stop the crack from propagating (hence the word stop hole), at least until a better repair could be made later. Unfortunately, later may be much later and recent studies (ref. Fatigue Technology, Inc., 150 Andover Park West, PO Box C-88388, Seattle, WA 98188 and Grumman Aerospace Corp., Bethpage, NY 11714) have indicated that repropagation may occur from a properly placed stop hole after an initial period of time. If the stop hole failed to completely clean out the crack, this initial time would be even shorter. The point is that stop holes must completely surround the point(s) of the crack and then only relied on for a short period of time if continuing high stress use of the part is anticipated.

Verification of crack removal can be accomplished by various means. Among the most common is *penetrant inspection*, usually used for aluminum parts, but applicable to other materials as well. Several proprietary techniques exist and the applicable specification is MIL-I-6866. These techniques utilize a high-penetration dye or photographic emulsion that deposits within or across any remaining crack and can be visually observed after special treatment. Another technique is governed by MIL-I-6868, *magnetic particle inspection*, and is extensively used on parts that can be magnetized such as many steels. Aluminum cannot be inspected for cracks in this manner. The parts to be inspected are magnetized and then hosed down with a slurry containing iron fillings. These particles are attracted to any crack, since the opposite sides of the crack will set up a magnetic field. After a suitable rinse, the remaining particles will become visible under ultraviolet light.

Several electronic techniques (such as *eddy current inspection*) are available for detecting cracks. These techniques often can detect cracks below the surface of a part, but not extending to the surface. They also have the capability of determining the depth of the crack from the surface. The responsibility for the use of these techniques rests with the inspection or quality control department, but the MRB engineer must be aware of them and sometimes request their use.

Additional mechanical attention beyond stop holing is required if the adverse effects of cracks are to be eliminated. If the part containing the crack is to be left in place, with or without any required reinforcement, the faces of the crack surfaces must be smoothly blended if the crack is completely through the part, or the full depth of the crack must be completely removed if the crack is only partway through the part. For cracks completely through the thickness of the part, the crack line must be slotted out with a tool (such as a router bit) and the walls of the slot blended smoothly with a suitable abrasive, or possibly blended and then shot peened. For cracks only partway through the thickness of the part, all evidence of the crack must be completely removed by smooth blending, and the complete removal verified by a suitable inspection technique such as penetrant or magnetic particle inspection. Cracks should never be left in place. The smoothness of the blendout should be at least as fine as that of the surrounding area and preferably smoother. Additionally, the blended surface should have any missing or removed surface coatings such as alodine, anodize (treatments applicable to aluminum), plating, paint-type primers and coatings reapplied. Any restorative surface treatments such as shot peening must be done before the reapplication of finishes.

Electrical Bonding

The requirement for electrical bonding, often referred to by the single word *bonding*, is sometimes confused with adhesive bonding (also often called bonding). Electrical bonding is accomplished by ensuring that a path for electrical conductivity exists between adjoining metallic parts to eliminate the possibility of static charge buildup, electric shocks, sparks, and interference with on-board electrical or electronic equipment. To accomplish this, the adjoining parts required to be bonded must be permanently joined, bare metal to bare metal, either by metallurgical techniques (such as welding) or mechanically through the use of permanent fasteners. If the parts are removable, or intended to be moved during use (such as doors), they require that a flexible metallic bonding cable or jumper be installed between them. The important requirement is that the metallic surface of both the bonding jumper end connections and the parts against which it is to be installed are completely free of finishes, oils, and so forth, before reinstallation. The only exception would be the presence of an electrically conductive coating as part of the basic design. If in doubt, an electrical continuity check is required.

Sealing of Fasteners to Resist Moisture Intrusion

In some instances of repair activity, the MRB engineer may become aware of the possibility of moisture intrusion into the area underlying the repair. This may be especially evident with assemblies returned from service and exhibiting evidence of corrosion upon disassembly. Once the ill effects of the corrosion have been removed and the repair design is well under way, the engineer must consider the need to leak-proof the relevant fasteners, possibly some of the remaining original fasteners as well as the newly installed repair fasteners. The sealing of these fasteners can be accomplished in several ways. One method is installation of an elastomeric O ring on the shank of the fastener and against either face of the part through which moisture may be expected to intrude.

A suitable recess must be provided either in the fastener itself or the part against which the O ring is to be compressed. The recess must not be so large, however, that the strengths of the recessed parts are compromised or the O ring is not compressed enough to satisfactorily resist the fluid pressure anticipated.

Another similar device is a sealing washer, such as the NAS1523, which is a flat washer with a built-in rubber sealing ring around its inside diameter.

It is recommended that both the O ring and the sealing washer be used with fasteners that are to be permanently installed. When used with removable fasteners, they can become displaced and lost. The repair technician may not realize the ring or washer was required, or a replacement may not be readily available.

Other less sophisticated sealing methods require the use of wet primer or a special sealant when installing the fasteners to be sealed. Either wet primer or sealant can be used to coat the hole bore and the shank of the fastener immediately prior to the insertion of the fastener into the hole. For this technique to work properly, the fastener must have a snug fit in the hole; no sloppy fits allowed. The sealant, generally much more viscous or sticky than the wet primer, has the added advantage in that it also can be applied as a fillet or topcoat around and on the head and/or the tail of the fastener to provide an additional barrier against leakage.

Machined Part Fabrication

Machined part processing takes many forms, including the removal of metal from the block of material (called the *billet*, *bar*, *plate*, or *rod*) by not only mechanical means, but by electrical discharge machining, electrochemical grinding, chemical milling, and electrochemical machining. Probably the most used is mechanical machining, the removal of controlled areas of the stock piece by cutting or grinding. Grinding is accomplished by the use of special abrasive tools of many shapes and sizes and actually erodes away the material. The grindstone that sharpens a chisel also can remove metal other than to produce the usual sharp edge. Careful attention must be given to the rotational speed of the grinding wheel, the speed of the part past the path of the wheel (or vice versa), the depth of material removal, the heat buildup developed, wear of the grinding wheel, and any necessary lubricants.

The mechanical removal of metal or other materials (since machined parts also may be nonmetallic) by other than grinding is accomplished by a variety of cutting tools, each possessing one or more cutting edges moved against or into the part by carefully controlled means. These tools vary from the common drill bit or reamer (a specialized form of drill having its primary cutting action along the length of its shank rather than at its tip) to the milling cutter, more like a rotating drum with cutting edges on its end, its body, or both. Metal cutting saws also are used for certain tasks, and machines called shapers and planers use specially designed metal cutters, as do routers. The thing most machined part material removal tools have in common is the fact that either they themselves rotate about a central tool axis, or the part being machined rotates about a fixed tool. In many cases, both the tool and the part move with respect to each other, controlled either by a human operator standing in place at the work site, or by numerical or computer-controlled electronic means. Machining is highly specialized and the more exposure to this field the structural MRB engineer can experience the better, insofar as his or her ability to work with machined part problems and solutions is concerned.

As previously mentioned, the same variables involved with the grinding of machined parts apply to mechanical machining. Failure to follow the requirements may result in a defective part, experiencing not only the common excessive removal of material, but the occasional generation of excessive heat caused by inadequate lubricant, dull tools, or incorrect speeds and feeds. The appearance of the machined surface often is a giveaway to the precision of the operation, but the MRB engineer must realize that different types of machining operations cause different surface textures and roughness (see p. 44). Mismatches along the direction of cutter travel due to uneven but parallel cutter passes also are common and often are allowed by specification, but only up to a defined limit.

Other types of surface conditions produced by improper tool use are cutter dwell and cutter chatter, both producing uneven surfaces. These often are caused by operator error or machine malfunction, but there are design criteria that the machine part designer should follow to minimize tendencies toward error. These criteria include the selection of suitable fillet and corner radii; depths of cut for features such as slots; provisions for clearance for the tool spindle or arbor; curved surfaces versus a series of varying straight line surfaces; chamfers instead of radii; inside corners; extreme angles; and (among the most important of all) the selection of appropriate tolerances, not only those necessary to provide a precisely fitting part but those allowable to permit the manufacture of an economically affordable part.

Welding—Seek Technical Assistance

Welding often can be used as a method of repair and the MRB engineer should become familiar with the types commonly used within his facility. One of the best ways to gain familiarity is to check the manufacturing process requirements associated with those production parts requiring welding as part of their fabrication. Obtain a copy of the required welding specifications and study the contents, making notes along the way. In this way, the MRB engineer can build up a personal library of valuable and applicable knowledge, not only for welding, but for other techniques as well. Many types of welding exist, from electrical arc welding, spot, and seam welding through tungsten inert gas welding to plasma arc and electron beam welding. Each technique has its specific restrictions and limitations, and each type of metal is restricted to particular types of welding. Different metals and alloys vary in their degree of weldability; the attainable weld strengths also vary.

Since all welding generates heat, this may be a problem. The occurrence of a heat-affected area to either side of the completed weld requires evaluation. This is especially so if the part is heat treated before welding, as an annealed or softened area along the weld occurs. If the part can be welded before heat treat and then heat treated, the higher attainable strength may be realized throughout the weld zone. Unfortunately, most physical damage to welded parts occurs after heat treatment.

Some drawings may state that annealed areas after weld are permissible. This can be most helpful to the MRB engineer considering a weld repair. Additional requirements may exist following the welding operation, such as finish grinding to reduce the height of the weld beads or stress relief, an intermediate temperature thermal treatment to remove the undesirable residual (locked in) stresses caused by the welding operation. Some companies employ welding design specialists or welding engineers; often their approval and signature is required on any document specifying a welding operation. Their assistance in the preparation of an MRB disposition calling for a weld repair may be invaluable and their advice in such matters should be requested, unless the MRB engineer is a weld specialist, well-qualified in the nuances of welding technology.

Appendix

Note: The following is excerpted from MIL-STD-1520C

Contents

1.   Scope

1.1 *Purpose.* This standard sets forth the requirements for a cost-effective corrective action and disposition system for nonconforming material. It defines requirements relative to the interface between the contractor and the contract administration office on nonconforming material. This standard sets forth the DOD contracting activity requirements for a properly constituted material review board. The primary purposes of the corrective action and disposition system are to identify and correct causes of nonconformances, prevent the recurrence of wasteful nonconforming material, reduce the cost of manufacturing inefficiency, and foster quality and productivity improvement.

3.5 *Material review board (MRB).* A board consisting of representatives of contractor departments necessary to review, evaluate, and determine or recommend disposition of nonconforming materials referred to it.

3.6 *Nonconformance.* The failure of a characteristic to conform to the requirements specified in the contract, drawings, specifications, or other approved product description.

3.6.1 *Minor nonconformance.* A nonconformance which does not adversely affect any of the following:
    a.   Health or safety
    b.   Performance
    c.   Interchangeability, reliability, or maintainability
    d.   Effective use or operation
    e.   Weight or appearance (when a factor)
Note: Multiple minor nonconformances, when considered collectively, may raise the category to a major/critical nonconformance.

3.6.2 *Major/critical nonconformance.* A nonconformance other than minor that cannot be completely eliminated by rework or reduced to a minor nonconformance by repair.
Note: Where a classification of defects exists, minor defects are minor nonconformances. Major and critical defects which cannot be completely eliminated by rework or reduced to a minor nonconformance by repair are major/critical nonconformance.

3.7 *Nonconforming material.* Any item, part, supplies, or product containing one or more nonconformances.

3.8 *Occurrence.* The first time a nonconformance is detected on a specific characteristic or a part or process. All nonconformances attributed to the same cause and identified before the date, item, unit, lot number or other commitment for effective corrective action also are considered occurrences.

3.9 *Recurrence.* A repeat of a nonconformance other than provided for in paragraph 3.8.

3.10 *Preliminary review (PR).* An evaluation by contractor-appointed quality personnel, assisted by other personnel as required, to determine the disposition of nonconforming material after its initial discovery and prior to referral to the MRB. PR may result in an authorized disposition of the nonconforming material without referral to the MRB for final disposition.

3.11 *Quality improvement project (QIP).* An activity chartered and monitored by the CAB (or other contractor group senior to the CAB) to investigate technology, methods, and procedures, which is aimed at finding more efficient and effective means of carrying out contractual responsibilities with the objective of enhancing quality and productivity.

3.12 *Repair.* A procedure which reduces but does not completely eliminate a nonconformance and has been reviewed and concurred in by the MRB and approved for use by the government. The purpose of the repair is to reduce the effect of the nonconformance. Repair is distinguished from rework in that the characteristic after repair still does not completely conform to the applicable drawings, specifications, or contract requirements. Except for SRPs (see paragraph 3.15), proposed repairs approved by the government are authorized for use on a one-time basis only.

3.13 *Rework.* A procedure applied to a nonconformance that will completely eliminate it and result in a characteristic that conforms completely to the drawings, specifications, or contract requirements.

3.14 *Scrap.* Nonconforming material that is not usable for its intended purpose and which cannot be economically reworked or cannot be repaired in a manner acceptable to the government.

3.15 *Standard repair procedure (SRP).* A documented technique for repair of a type of nonconformance which has been demonstrated to be an adequate and cost-effective method of repair when properly applied. SRPs are developed by the contractor, reviewed and concurred in by the MRB, and approved by the government for recurrent use under defined conditions. Defined conditions shall include an expiration date or a finite limit on the number of applications, or both.

3.16 *Statistical process control (SPC).* SPC is a methodology used to measure the average and variability of any given characteristic within a contractor area, department, part, or process, including but not limited to machine shop, bonding process, heat treat, and assembly. SPC techniques include control charts and control limits. Properly implemented, SPC offers the ability to improve manufacturing yield and lower production, inspection, and nonconformance costs.

3.17 *Supplier.* The terms subcontractor, supplier, vendor, seller, or any other term used to identify the source from which the prime contractor obtains support, are considered to be synonymous for the purpose of this standard.

3.18 *Use-as-is.* A disposition of material with one or more minor nonconformances determined to be usable for its intended purpose in its existing condition.

3.19 *Definitions of acronyms used in this standard.* Acronyms used in this standard are listed and defined as follows:

    a. CAB – Corrective action board
    b. DOD – Department of defense
    c. FAR – Federal acquisition regulation
    d. MRB – Material review board

4.1 *Corrective action and disposition system.* The contractor shall establish and maintain a system which shall identify, segregate (or control if segration is not practical), and properly dispose of nonconforming material and shall ensure that cost-effective, positive corrective action is taken to prevent, minimize, or eliminate nonconformances. The system shall work toward continual improvement of quality and productivity through the initiation and monitoring of QIPs.

4.2 *Statistical process control (SPC).* SPC techniques, including control limits and control charts, shall be used when appropriate. Control limits must be established statistically or by other methods acceptable to the government and be based upon the documented history of the process capability.

4.2.1 *Control limit standards.* Nonconformances due to chance causes can occur that may not warrant individual corrective action. As an alternative to individual corrective action, the contractor may develop and recommend to the government the use of a standard(s) to control such nonconformances. Contractor-recommended standards shall specify the control limits at which corrective action must be taken: describe criteria for determining the control limits: and provide for the accumulation and maintenance of data for monitoring processes and obtaining corrective actions as dictated by collective analyses, trend reviews, or other means.

4.3 *Quality improvement.* The contractor shall institute actions to prevent nonconformances and initiate QIPs throughout the contractor's organizations. The contractor shall assign organizational elements, teams, or individuals to investigate technology, methods, and procedures to increase efficiency and conformance to requirements. The contractor shall monitor the QIP progress toward established goals at regular intervals. The requirements of this paragraph shall be the responsibility of the CAB or, at the discretion of the contractor, of a contractor group senior to the CAB.

4.4 *Contractor's written procedures.* The requirements of this standard shall be implemented by the contractor through the preparation, publication, and maintenance of detailed written procedures. The contractor shall identify personnel-appointed PR authority and those to act on the MRB and CAB, and shall indicate in the procedures the scope or extent of their authority.

The contractor's procedures also shall indicate the manner in which documentation is maintained.

4.5 *Material review board (MRB).* The MRB shall be chaired by a representative of the contractor's quality organization and shall include, as required, personnel representing other contractor organizations necessary to determine appropriate disposition of nonconforming material. As a minimum, the MRB shall include the chairman and a representative of the contractor's engineering organization responsible for product design. MRB members shall be selected on the basis of their technical competence. MRB members may call upon other contractor personnel for technical advice. If warranted by the volume of nonconforming material or the diversity of work operations, more than one MRB may be established.

4.5.1 *MRB authority and responsibilities.*

a. The MRB shall investigate, in a timely manner, all nonconforming material (except material previously disposed of in PR authorized in paragraphs 5.2 a, b, c, or d) in sufficient depth to determine proper disposition.

b. The MRB shall review and concur in all proposed use-as-is and repair dispositions prior to submission to the government for approval.

c. The MRB shall review and concur in all proposed SRPs prior to submission to the government for approval for recurrent use under defined conditions.

d. A written engineering analysis shall accompany proposed use-as-is and repair (excluding SRP) dispositions if requested by the government. The MRB shall ensure that the government is kept informed of its investigation and deliberations on these potential dispositions so that the government may act upon the MRB recommendations in a timely manner.

e. The MRB shall dispose of nonconforming material in accordance with paragraph 5.3.

4.6 *Corrective action board (CAB).* The CAB shall ensure that an effective corrective action system is functioning throughout the contractor's organization. This function shall be performed through review and analysis of nonconformance data. The CAB shall ensure that records of causes of nonconformances, trends, and individual causes acted upon are maintained and that individual records and summaries of actions taken are prepared. If warranted by the diversity of work operations, more than one CAB may be established.

4.6.1 *CAB authority and responsibility.*

a. The CAB shall have authority to ensure implementation of corrective actions throughout the contractor's organization. The corrective actions shall extend to all contractor operations affecting product quality.

b. The CAB shall have the authority to require investigations and studies by other contractor organizations necessary to define essential corrective actions which will result in reducing nonconformance costs and reducing the amount of nonconformances.

c. The CAB shall ensure that documentation required by paragraphs 5.7, 5.7.1, 5.7.2, 5.7.3, 5.7.4, and 5.8 is maintained.

d. The CAB shall ensure that summary data of nonconformances and associated costs are analyzed and areas of high potential payoff, adverse trends, exceeding control limits, or out-of-control recurrences of nonconformances are thoroughly investigated to identify appropriate corrective actions and to identify potential QIPs.

e. The CAB is responsible for ensuring that follow-up systems are maintained to ensure that timely and effective corrective actions are taken.

f. The CAB shall ensure that reviews of nonconformance data and PR and MRB disposition decisions are conducted periodically to determine that PR and MRB actions are effective and in compliance with the requirements of this standard.

g. When control limit techniques are used and analysis of cumulative data for an applicable nonconformance reveals that the established limits are being or will be exceeded, the CAB shall ensure that a process evaluation is accomplished and that specific corrective actions are taken to bring the process back into acceptable limits.

h. When corrective action is required due to inadequate process controls or control limit techniques and until such time as it has been demonstrated that the corrective action has been effective, the CAB shall ensure that the contractor documents nonconformances and monitors: yield requirement development, documentation, and evaluation; the process control system for compliance; process improvement activity as it relates to trends; and recurrences of nonconformances.

i. The CAB shall be responsible for the initiation and monitoring of QIPs unless this function has been assigned by the contractor to a group senior to the CAB.

4.7 *Government rights.* The government reserves the right to: review all contractor procedures developed to implement this standard; disapprove the procedures if they do not accomplish their objectives; observe PR, MRB, CAB, and QIP activities; and review documents or other data required by this standard. Acceptance or rejection of nonconforming material presented to the government is the sole prerogative of the government. Acceptance of nonconforming material by the government may involve a monetary adjustment or other consideration. The right of government disapproval specifically applies, but is not limited to the following:

a. Procedures, activities, organization, and reports of PR, the MRB, and the CAB.

b. Contractor standards establishing control limits.

c. Contractor-proposed repair procedures, including SRP expiration dates, limits, and extensions.

d. Records and analyses of nonconformances and corrective actions related to those nonconformances.

e. The right to withdraw approval of previously approved SRPs.

f. MRB and CAB members and personnel appointed PR authority at the time of selection or anytime thereafter.

5.1 *Identification and segregation of nonconforming material.* When material is found to be nonconforming, nonconforming items shall be conspicuously marked or tagged (or otherwise identified if marking or tagging is not practical) and positively controlled to preclude its unauthorized use in production. Nonconforming material to be submitted to the MRB shall be moved to a controlled area designated for storage of nonconforming material unless impractical due to size, configuration, environmental requirements, or other conditions authorized by the government. The designated area shall be protected to preclude unauthorized removal of nonconforming material.

5.2 *PR disposition.* When material is initially found to be nonconforming, it shall be examined by contractor-appointed quality personnel, assisted by other contractor personnel if necessary, to determine if the nonconformance

a.   Requires scrapping of the material because it is obviously unfit for use and cannot be economically reworked or repaired.
b.   Can be eliminated by rework.
c.   Requires return of the material to the supplier.
d.   Can be repaired using SRPs which have been concurred in by the MRB and approved by the government.
e.   Meets none of the above criteria and shall be referred to the MRB.

PR action does not negate the requirement for identification, documentation, and corrective action associated with nonconformances. It does recognize that some nonconformances do not warrant referral to the MRB and can be handled more economically at the location of initial detection.

5.3 *MRB disposition.* All nonconforming material not disposed of in PR shall be disposed of by an MRB decision to

a.   Scrap.
b.   Rework.
c.   Return to supplier.
d.   Repair by an approved SRP.
e.   Recommend to the government for repair by other than an SRP.
f.   Recommend to the government for use-as-is.
g.   Request a waiver from the contracting officer.

5.4 *Use-as-is disposition.* Requirements pertaining to use-as-is dispositions are as follows:

a. All use-as-is dispositions must be approved by the government.
b. Until the usc-as-is disposition has been approved, the nonconforming material shall not be further processed nor used without prior government authorization, or unless controlled by methods approved by the government.
c. All use-as-is dispositions shall include a determination of the appropriateness of a documentation change and the method for accomplishing any recommended change (in other words, design change, changes to technical documentation including drawings, specifications, and technical orders, or recommended changes to government specifications).

5.5 *Repair disposition.* Requirements pertaining to repair dispositions are as follows:

a. SRPs shall be submitted to the government for approval prior to implementing the SRP.
b. Proposed repair methods (other than previously approved SRPs) shall be submitted to the government for approval prior to accomplishing the repair action.
c. The government act of approving the repair technique does not compromise the government's right to reject the material after completion of the repair. Use of all repair procedures is at the contractor's risk.
d. Prior to any repair disposition decision, a judgment shall be made by the contractor that the repair will be cost-effective relative to other disposition alternatives.
e. Instructions for reprocessing of material after completion of repair and before its release shall be included in the SRP or other repair procedure. These procedures shall include the requirement for contractor inspection and test.
f. The contractor shall maintain records detailing the dates of use and number of applications of SRPs.

g. The contractor shall review SRPs periodically to ensure that they are complete, up-to-date relative to current process capability and state-of-the-art, and are being properly applied under the conditions defined for their use.

h. Nonconforming material to which an SRP has been satisfactorily applied is subject to government inspection when specified in the SRP or as otherwise directed by the government. All other repaired material shall not be further processed nor used without prior government authorization or unless controlled by methods approved by the government.

5.6 *Scrapped material.* Scrapped material shall be conspicuously identified and controlled to preclude its subsequent use in a contract item unless approved by the government.

5.7 *Nonconforming material documentation.* The contractor system shall maintain records for all nonconforming material, dispositions, assignable causes, corrective actions, and effectiveness of corrective actions. These records shall be organized to permit efficient retrieval for summarization required by paragraph 5.8, knowledge of previous dispositions, and corrective action monitoring. The contractor shall ensure that documentation of nonconformances includes the following:

a. Contract number.

b. Initiator of the document.

c. Date of the initiation.

d. Identification of the document for traceability purposes.

e. Specific identification (for example, part number, name, national stock number) of the nonconforming material.

f. Quantity of items involved.

g. Number of occurrences.

h. The place in the manufacturing process where the nonconformance was detected.

i. A detailed description of the nonconformance.

j. Identification of the affected specification, drawing, or other document.

k. A description of the cause(s).

l. Disposition of the nonconforming item (return to supplier, rework, use of an SRP, scrap, or refer to MRB).

m. Identification of personnel responsible for making the disposition decision.

5.7.1 *Additional documentation for MRB items.* If nonconforming material is referred to the MRB for disposition, the MRB shall add the following information to the documentation:

a. Reference to or attachment of the written engineering analysis when performed.

b. Final disposition of the nonconforming items.

c. Signature (or personal identification stamp) of disposition authorities.

5.7.2 *Additional documentation for corrective action.* If corrective action is required on an individual nonconformance, the following information shall be recorded:

a. An analysis of the recorded cause(s) and identification of the true (or root) cause.

b. The actions taken (or planned) to correct the cause(s) of the non-conformance and thereby preclude recurrence.

c. Identification of the individual(s) and contractor functional area(s) responsible for taking the corrective action.

d. Date, serial number, or lot number when corrective action will be completed or is estimated to be completed.

5.7.3 *Recurring nonconformances.* If corrective action is not warranted on an individual nonconformance, but collective or trend analyses of recurrences of the nonconformance indicate that the process is not within acceptable limits and corrective action is necessary, the contractor shall document the information required by paragraph 5.7.2. This information need not be included on the individual nonconformance records.

5.7.4 *Nonconformance costs.* The contractor shall determine and record the costs associated with nonconformances. The objective of generating this cost data is to provide current and trend data to be used by the contractor in determining the need for and effectiveness of corrective action. The resultant cost data shall serve as a basis for necessary CAB and QIP action when appropriate. Nonconformance cost summaries shall, upon request, be furnished to the government. The cost collection shall consist of scrap, rework, repair, use-as-is, and return to supplier costs, plus other costs as determined appropriate by the contractor.

5.8 *Minimum data summarization requirements.*
Nonconformance data shall be recorded to enable summarization of the quantity of nonconforming items, number of recurrences, cause determinations, corrective actions, dispositions, and nonconformance costs as described in paragraph 5.7.4. Nonconformance data shall be used by the CAB to determine the need for and effectiveness of corrective action. The format of the data and the frequency of preparation shall be at the discretion of the contractor but in no case shall the preparation be less frequent than quarterly. As a minimum, the following data shall be included:

    a.   Quantity of nonconforming items.
    b.   Number and type of nonconformances.
    c.   Number and type of dispositions.
    d.   Cause determinations.
    e.   Type of corrective actions and status.
    f.   Delinquent corrective actions.
    g.   Nonconformance costs.
    h.   Trend information and analysis thereof.

5.9 *Control of material review and disposition system at suppliers.* The prime contractor has the option to delegate to suppliers the authority for material review and disposition of nonconforming material. If the prime contractor elects to delegate such authority, the procedures of this standard shall apply either in full, or as appropriately tailored, to the suppliers. Tailored requirements applied to supplier shall be in consonance with the requirements of this standard and must be acceptable to the government. Furthermore, the authority to present nonconforming material to the government for approval of recommended dispositions is limited to the prime contractor's MRB unless specific authority has been delegated to the government agency having contract administration responsibility for the prime contract. The prime contractor shall review and approve material review and disposition systems of suppliers.

5.9.1 *Corrective action at supplier facilities.* Supplier organizations shall be notified of material nonconformances and the requirement, when necessary, for corrective action. The contractor shall perform follow-up review of the corrective action taken by suppliers.

5.9.2 *Records of nonconforming material received from suppliers.* The contractor shall maintain a record of any nonconforming material received from each supplier. This information shall be used in the contractor's vendor or supplier rating system.

5.10 *Audits.* The contractor shall periodically audit, or have audited, the corrective action and disposition system for nonconforming material (both in-house and at suppliers where appropriate) for compliance with the requirements of this standard and to ensure effectiveness. If an audit is conducted by a part other than the prime contractor, the contractor should notify the government and remain primarily responsible for that performance.

References

1. "Advanced Composite Repair Guide," Wright-Patterson Air Force Base, by the Northrup Corporation.
2. ASTM-G-47, "Test Method for Determining Susceptibility to Stress Corrosion Cracking of High Strength Aluminum Alloy Products."
3. MIL-D-8706, "Data and Test Engineering: Contract Requirements for Aircraft Weapons Systems" (Superseded by SD-8706).
4. MIL-H-6088, "Heat Treatment of Aluminum Alloys."
5. MIL-H-6875, "Heat Treatment of Steels (Aircraft Practice, Process for)."
6. MIL-HDBK-5E, "Metallic Materials and Elements for Aerospace Vehicle Structures."
7. MIL-HDBK-17B, "Polymer Matrix Composites."
8. MIL-HDBK-23, "Analysis and Design for Sandwich Construction."
9. MIL-I-6866, "Inspection, Liquid Penetrant."
10. MIL-I-6868, "Inspection Process, Magnetic Particle."
11. MIL-I-8500, "Interchangeability and Replaceability of Component Parts for Aerospace Vehicles."
12. MIL-Q-9858, "Quality Program Requirements."
13. MIL-R-46082, "Retaining Compounds Single Component, Anaerobic."
14. MIL-S-7742, "Screw Threads, Standard, Optimum Selected Series: General Specification for."
15. MIL-S-8879, "Screw Threads, Controlled Radius Root with Increased Minor Diameter, General Specification for."
16. MIL-STD-105E, "Sampling Procedures and Tables for Inspection by Attributes."
17. MIL-STD-414, "Sampling Procedures and Tables for Inspection by Variables for Percent Defective."
18. MIL-STD-481B, "Configuration Control-Engineering Changes (Short Form) Deviations and Waivers."

19. MIL-STD-1520C, "Corrective Action and Disposition System for Nonconforming Material."
20. MIL-STD-1537, "Electrical Conductivity Test for Verification of Heat Treatment of Aluminum Alloys, Eddy Current Method."
21. MS9226, "Wire-Steel, Corrosion and Heat Resistant, Safety, 1800° F."
22. MS20995, "Wire, Safety or Lock."
23. MS24665, "Pin, Cotter (Split)."
24. MS33633, "Inserts, Screw Threaded, Design and Usage Limitation for."
25. NAS1523, "Packing with Retainer."
26. Frocht, M. M., *Photoelasticity*, Vol. II, New York: John Wiley and Sons, 1948.
27. Seely, F., *Resistance of Materials*, New York: John Wiley and Sons, 1935.
28. SD-24, "General Specification for Design and Construction of Aircraft Weapon Systems."
29. Peterson, R. E., *Stress Concentration Design Factors*, New York: John Wiley and Sons, 1953.
30. Timoshenko, S., *Theory of Elastic Stability*, New York: McGraw-Hill.
31. Neuber, Heinz, *Theory of Notch Stresses: Principles for Exact Stress Calculation*, Ann Arbor, MI: J. W. Edwards, 1946.

Index